T0210726

Perceptions of Climate Change from North India

Perceptions of Climate Change from North India: An Ethnographic Account explores local perceptions of climate change through ethnographic encounters with the men and women who live on the front line of climate change in the lower Himalayas.

From data collected over the course of a year in a small village in an eco-sensitive zone in North India, this book presents an ethnographic account of local responses to climate change, resource management and indigenous environmental knowledge. Aase J. Kvanneid's observations cast light on the precarious reality of climate change in this region and bring to the fore issues such as access to water, Non Governmental Organisation's intervention and climate information for farmers. In doing so, she also explores classic topics in the study of rural India, including ritual, gender, social hierarchy and political economy. Overall, this book shows how the cause and effect of climate change is perceived by those who have the most to lose and explores how the impact of climate change is being dealt with on a local and a global scale.

This book will be of great interest to students and scholars of the anthropology of climate change, environmental sociology and rural development.

Aase J. Kvanneid is an anthropologist, currently working as an associate professor of Global Development Studies at the University of Agder and as a postdoctoral researcher at the Department of Cultural Studies at the University of Oslo. Her main areas of research are the societal aspects of environmental and climate change, and she is currently researching the empirical embeddedness of sustainability and transcendental visions in Asia.

Routledge Advances in Climate Change Research

For more information about this series, please visit: https://www.routledge.com/Routledge-Advances-in-Climate-Change-Research/book-series/RACCR

Perceptions of Climate Change from North India

An Ethnographic Account

Aase J. Kvanneid

First published 2021
by Routledge
2 Park Square, Milton Park, Abingdon, Oxon OX14 4RN

and by Routledge
52 Vanderbilt Avenue, New York, NY 10017

Routledge is an imprint of the Taylor & Francis Group, an informa business

British Library Cataloguing-in-Publication Data
A catalogue record for this book is available from the British Library

Library of Congress Cataloging-in-Publication Data
A catalog record has been requested for this book

ISBN: 978-0-367-42143-4 (hbk)
ISBN: 978-0-367-72419-1 (pbk)
ISBN: 978-0-367-82214-9 (ebk)

Typeset in Times New Roman
by codeMantra

To Aksel and Idunn

Contents

Figures

Acknowledgements

For all its weaknesses and flaws, this work is entirely mine. In the case of any misrepresentations, or any breaches of the trust bestowed in me, the responsibility lies on my shoulders only.

For all its strengths, this work has also been the result of the efforts of others. First and foremost, I wish to thank my family, including my extended family in Rani Mājri. Without you, there would be no book at all, and I am deeply grateful for your support and care. I would also like to thank the wise men and women who read and commented upon earlier versions of this text, providing insight along the way. I also want to thank the department of Social Anthropology at the University of Bergen, who funded the fieldwork upon which this book is based.

Note to the reader

The chapters of this book are structured around the six Indian seasons, which usually consist of two lunar months. The dates of the (solar and lunar) Hindu calendar are not fixed and vary with the timings of the sun and the moon, but they generally correspond to mid-month to mid-month of the Gregorian calendar. Both calendrical systems were used.

Hindi/Urdu/Pahari expressions have been transcribed and Romanized using John T. Platt's *Dictionary of Urdu, Classical Hindi and English* (2008) and/or dialogue with professor of South Asia studies Claus Peter Zoller, unless otherwise stated.

Some simplifications have occurred when, or if, they have established equivalents in roman characters, such as Śiva to Shiva, Panćāyat to Panchayat, and so on. Certain words have no settled equivalent, such as the Shivalik Hills, which is also spelled "Sivalik" and "Siwalik". As the "Shiva" in "Shivaliks" refers to the Hindu deity, I have chosen Shivalik throughout the text.

I have tried to clarify in what ways the practices I describe might diverge based on caste and gender, but it is important to keep in mind that I base my account largely, but not exclusively, on the practices of Rajput women. Lastly, most names of individuals and places are fully anonymized.

Prologue

In the wee morning hours, just as monsoon season begins, sounds and smells from the lives led in this small hillside village fill the warm air. A light chime from the goats in the courtyards tinkles in the air. A dog saunters around somewhere. The odour of cattle-dung fuses with the smell of coriander, grass and earth. The faint smell of firewood from the hearths suggests the presence of women making breakfast.

In one of the larger landholding houses of the village, a middle-aged Rajput woman is sitting alone on the earthen floor of a small kitchen. She is preparing breakfast for a joint family consisting of two elderly, six adults, four teenagers and three foreigners – an anthropologist, her husband and their baby. Her name is Nirmala, and she is the wife of Prakash; they are a middle-aged couple living in a joint family of twelve. Cross-legged on the earthen floor, Nirmala is rhythmically kneading the dough of the flat and round bread, the light wheat flour *phulkā*. By skilfully moving vessels and pots back and forth upon the hearth simultaneously, she re-heats the thicker yellow maize bread – the makkī rotī from yesterday's dinner – and makes the sweet and rich black buffalo-milk and water tea *(chaī)*. The fire is hot, and the air in the dark room is warm, but the earthen floor of the kitchen retains the coolness of night. Drowsy and cross-legged the women of the house, including the anthropologist, chew the bread still warm from the hearth, smeared with clarified butter, salt and chili. Suddenly, heavy rain falls from monsoon skies, a heavy, grey curtain over Rani Mājri. The pre-monsoon showers had been going on for a few weeks, and with damp sighs, people stay indoors to wait it out.

As quickly as it began, the rain halts, the skies clear and the sun breaks through. With my husband looking after my son, I too begin my morning routine: washing the laundry. I am able to use the water in the bathroom for washing, and I know this is a privilege. These water-logged days, the water in the small channel where women normally wash their laundry is hasty and brown with debris, leaves and twigs. Immersed in my own thoughts, I suddenly hear a deep rumble. At first, I think maybe it is thunder, but the sound is deeper, more muddled, and feels so immediate. Puzzled, I step outside the bathroom and see my husband with my son on his arm and Babu, a young boy visiting the family, leaning curiously over the rails of the flat roof above me.

"Did you hear that?" they exclaim and run down the stairs towards me as I turn to see most of the land outside our bedroom wall gone, collapsed down into a riverbed about 100 meters below. The earthen patio outside the annexed room we rented had been expanded over the years with bricks to make space for a couple of buffaloes and a shed, but now, large parts of the patio have slid down the gully. A little curious to see the damage, I walk carefully out on the ledge but immediately step back. What feels like seconds later, the rest of the brick wall follows down the hill. There is now but a brink of land left, a few meters wide, between the house and the gully.

Padma and Naveen, Prakash's two eldest daughters, arrive and take initial control over the situation. Ordering everyone else to stand back, they call on their mother to let people know and begin to carry the remains of valuable materials, like bricks and logs, stored to enlarge the patio, away from the edge. Soon, Prakash arrives from the fields, and with grave faces, the family clear the ledge together.

In the hours that follow, Prakash sits down with worry on his mind, discussing the costs of the landslide with a friend. Apparently, they had tried to enlarge the ledge by mounting a brick wall there a few years before, intending to expand the area to keep buffaloes. This appears to have been a bad investment as several lakhs (one lakh is a hundred thousand *Rupees*) have now "gone down the hill". His friend tries to cheer him up, but Prakash's face remains sullen. His mother then approaches him with a jug of water, which he takes with him outside, without a word. I do not find it suitable to follow him.

Out of concern over new landslides, my husband and I drive to our city apartment with our son the same evening. Our bedroom bordered the site of the damage and – still breastfeeding and changing soiled nappies at night – I did not feel like accepting Nirmala's offer to share her bed in the main house.

That evening Prakash calls me, telling me that another rainfall had caused another small landslide at the same site but that they reckon the worst is over. He says he wants me to come back, which I do. When I return, alone this time, the sense of drama and urgency from the day before has cooled off.

That night, I talk to Prakash's younger brother's wife, Orpita, and I tell her I am worried about sleeping in the bedroom, which borders the slide-site. She admits she has been worried too but not so much anymore as they gave Khwaja Pir, a local deity, sweet porridge (called *dahlia*) during a worship (*pūja*) whilst I was in the city. Orpita lifts the last, puffed bread from the fire and patiently hunches down next to me, whispering that the risk is still high. They gave Khwaja Pir sweet porridge and asked if Bhagvān could "look their way", so everything will probably turn out fine, but one cannot really be sure. "He does as he pleases", she says, tellingly, "the way he [Khwaja Pir] has done it now – with all the water that has taken the wall, he should not do it like that! If he becomes angry, he might bring the whole house with him! It can happen!"

I was unfamiliar with Khwaja Pir at that time and had only heard his name mentioned briefly once before, when my son was ill with diarrhoea, and Prakash gave him *kuhl* water to drink to soothe his upset tummy. We had not lived in the village for long then, and at that point I had been more upset to learn that they had given him *kuhl* water, which I myself had fallen quite ill from drinking, than interested in someone called Khwaja Pir having done something to the water, making it OK for a baby to drink. Now, I cared.

Introduction

The description in the prologue was of a minor landslide in the rainy season of 2013, drawn from my fieldnotes, recorded when researching ethnographically the perceptions of climate change in the village of Rani Mājri in North India. Rani Mājri is a small village, located in the lower Shivalik hills, where the terrain begins its slow rise into eventually becoming the Himalayan mountains. Geographically bordered to the north by a seasonal river that runs in the southernmost hollow and to the north-east by a government-protected forest (*jangal*) the village lies at the end of a small, partly paved, partly dirt road. Towards the west the village fields fan out like a peacock's tail, flowing out onto the alluvial plains below, until they meet with other fields, belonging to lower-lying villages. To the south, the road becomes a path, running across yet another small ridge, past a small hamlet and another westward fan of fields that belongs to another tiny settlement of Khot. The path ends here. Up until that landslide, I had lived permanently in the village for over six months, in a small, annexed room with my husband and our baby son. The fright made us leave the village ahead of schedule, and I continued the rest of my fieldwork by commuting to Rani Mājri from Chandigarh, a city located a few hours down on the plains. The landslide did not only change the way I executed my fieldwork, but it changed the way I *perceived* how the villagers related to a changing environment and climate. The response to this landslide thus accentuated the precariousness of the changing weather, unpredictable seasons and an exhausted environment as much as it did the trenchant insecurity in which people must make plans and choices in their daily lives.

When I set out to do fieldwork for this book, I was particularly interested in how people use and care about the forest, the land, the village – whatever they define their environment to be – and whether they see their environment as being affected by climate change. So, I began my fieldwork by mapping property rights, trying to understand who could use what resources and to what degree they were concerned about climate change and environmental decay. Temples and shrines, I found, were often built on what is called "common land", usually uncultivable land near the settlements and fields, along the paths that people walk anyway. In theory, this common land is

Figure 0.1 Case study location in India.
Source: Bruce Jones Design Inc. (2010).
Image showing the location of Rani Mājri on a map outlining India.

available to anyone, yet certain shrines belong to certain groups, governed by caste, gender or kinship. I left it at that, for quite some time. Months had to pass before I realised that land, and who might use it for what, is not just a question of politics, money, gender or caste – there are others too that need to be consulted, convinced or pleased too, if the land provides its riches for you to eat, sell or give away. This became apparent through the explanations and responses to two unexpected monsoon-related incidents during my stay, both of which were thought to be the result of a complex chain of inter-related factors. One, a massive flood in the state of Uttarakhand and the other, the minor local landslide described in the prologue. In the village, these incidents evoked not only concerns related to social, economic, political and ecological issues but also ritual.

The response to the flood in Uttarakhand made me realise that, for the people of Rani Mājri, the cause and effect of climate change and global warming were related to notions of deteriorating environments, caused by deteriorating relationships. Many saw the flood as being directly related to Shiva, one of Hinduism's largest and most powerful gods, whose abode is in the Shivalik hills and whose tresses name the riverine valleys in which the flood hit. Shiva, it was thought, was upset with mankind, and thus he unleashed environmental retribution upon his devotees. As people increasingly let money and individual success steer their choices, rather than care for one's closest relationships, the world filled with sin, consequently heating it into a sort of feverish unpredictability. This, known as global warming, would cause disturbance to those finely tuned seasonal rhythms that men have built their lives around. Thus, the anger of Shiva, the melting glaciers and the seasonal disturbances were all connected to other kinds of changes too – between the village and the market; between people in power; between generations, genders and castes; and between those deities with which they dwell.

As the climate of the globe changes relentlessly, and the totality of life depending on it changes with it, the people of Rani Mājri are faced with more uncertainty and more insecurity. These floods, landslides and the increased insecurity of living a good life in these hills have become a natural part of an unnatural history, where villagers such as those in Rani Mājri play a marginalised part. In Rani Mājri, as my ethnography will convey, I did not find a vocalised concern for the relationship between carbon emissions and the atmosphere. However, I never observed unawareness of climate change either. On the contrary, by being attentive to what people saw as climate change-related shifts, as well as to how they were met and mediated with politics, economy, hard work and ritual, I met a highly aware village population who were continuously acting and engaging to make and remake their lives, and the unknown future into a benevolent one. However, as the fate and future of rainfall, temperature and soil health was thought to be in hands of God, the unparalleled rate of construction and resource extraction was in the hands of men – and so were the unabashed consumption and lifestyle transformations that accompanied those changes. This became particularly clear when observing how people planned for important or potentially life-changing events, such as the forthcoming harvest, a costly project business proposal, a wedding or a pregnancy. Faced by uncertainty, people would, as families and individuals, consult religious calendars and priests, and consider their advice on par with that from their relatives, friends, the school principal or Indian Institute of Soil and Water Conservation (I.I.S.W.C.) scientists, who were seen as an authority on scholarly knowledge. They would also pay attention to news via the newspaper or television and generally pick up any other bit of information that could help them make the best of what was perceived to be an unknown and potentially precarious situation.

Climate change in Rani Mājri was caused, driven and sustained by humans, thus making its consequences essentially social. This indicates a rich and varied spectrum of agency that people in this area employ in the face of climate change, and they prove to be far from passive victims, existing according to the whims of fate, modernity, governmental planning or angry deities. To utilise and benefit from various schemes and projects initiated by developers concerned about climate change relate to intricate relations, such as between the concrete individual or group, and the structures of global capitalism, contemporary Indian politics, caste essences, fierce gods and possessive spirits. "Sustainable development", in all its guises, is thus not for everyone. This also leads me to the core of the issue that I aim to address in this book: that the *scientific concept* of climate change reinforces an obscuration of the manifold perceptions of the *whys* and the *hows* that fuel and escalate those very changes.

The first three chapters show how government projects and schemes to aid Sustainable Development, now underscored and stressed in the time of climate change, unfold. As climate change and associated environmental deterioration are undeniably major issues in the Shivalik Hills, state and privately funded agencies, many of which have the best intentions, try to develop, aid and assist the villagers in navigating an unknown future. However, as the precarious situation enables larger and more extensive measures taken to improve crop yield, sanitation and infrastructure, state agencies and the discourse that guides and aids them, could very well be solidifying local structures of power. The apparent unawareness of the scientific concepts explaining the changes experienced, in turn, has served to legitimise certain policies, state-action plans and practices that at best corroborate certain stereotypes (e.g., the "backward" villager – a part of everyday speech – defined by Gupta (2005:423) as lacking "a culture of learning") and at worst reinforce certain social practices of marginalisation (such as effectively cutting the landless off from the local water irrigation structure). Chapter 1 explicitly addresses local principles of social differentiation and how they might cause schemes and projects to benefit some and disadvantage others. The next three chapters question what constitutes the "right kind of awareness" of the Sustainable Development discourse in North India and investigates its consequences. These chapters explore local perceptions of climate change and suggest that the perception of both the environment and climate is richer, more complex and far more dynamic than the structures of governance allow for. The final chapter draws upon previous ones to discuss the potential consequences of climate change discourse in North India as well as what different perceptions might offer to its mitigation. The last chapter asks whether climate change, in affecting how the environment is governed in Rani Mājri, might influence people's behaviour and thus their sense of self in the world. This introduction, however, sets the scene.

Climate change in India

At the time of writing India has become the fastest-growing and third-largest economy in the world (World Bank 2019; I.B.E.F. 2020). Simultaneously, it is also the second-most unequal country in the world (Shorrocks et al. 2016), where, despite a growing middle class, an estimated 600,000–700,000 children below the age of five die from deaths associated with malnutrition every year; every second Indian woman is anaemic; and millions more experience acute starvation and deprivations in health, schooling and sanitation whilst dealing with the severe issues of polluted water, soil and air (Swaminathan et al. 2019; UNICEF 2019). Increasing food costs, vigorous resource competition and persisting issues with poverty and inequality related to caste, gender, religion and nationality persist (Chansel and Piketty 2019). It is argued that climate change will exacerbate all these issues (Cruz et al. 2007; Hijioka et al. 2014; World Bank 2015; World Wildlife Fund 2017).

Mitigative measures of climate change in India have increasingly been on the political agenda of international relations and trade, affecting both domestic and foreign relations (Dubash 2013). For India in general, and the Himalayas and Shivalik Hills in particular, climate change is a wicked and serious problem (Cruz et al. 2007; Hijioka et al. 2014). Several research institutions have underscored the fragile state of the hills, as well as their population's vulnerability to climate change (Himalayan Climate and Water Atlas 2015). In a worst-case scenario, the latest United Nations International Panel on Climate Change (I.P.C.C.) report of 2019 predicts, 80% of the snow and ice in the Himalayas will melt, causing severe loss of drinking and irrigation water for millions of small-scale farmers. It will also add mercury and other harmful substances to the melted glacier water, and it is calculated that the financial losses from floods and mass movements have already mounted to US$45 billion in the Hindu-Kush Himalayan region over the period 1985–2014 (I.P.C.C. 2019).

Through the I.P.C.C. reports loom a sense of urgency and haste. They call for action now, before it is too late, and imply not only that Indian politicians must adhere to guidelines but that people must be made aware in order to be able to adapt to the changes that are occurring. As politics might prohibit, restrict or create incentives for industry, businesses and public institutions, the government emphasises that climate change adaption and mitigation cannot happen in India without the commitment of the millions of Indians who live in rural, Agricultural villages (Government of India 2011). The I.P.C.C. further states that people here are the "most vulnerable to the potential health impacts of climate change" in India (Wish 2010:2).

A rather unified scientific community has thus been vocalising the necessity of the state maintaining sustainable agriculture in the region and saving the livelihoods of people living off the land in the hills and in the plains regions.

The iconic position of the Himalayas and its villages as a site in need of international and national proactive measures was set in the media tumult that followed the I.P.C.C. report from Cruz et al. (2007). The report claimed that environmental and economic scenarios for India were grim if climate change was to continue unabated and postulated a climate change-affected future, with less predictable monsoon rains, the glaciers of the Himalayan icecaps melting rapidly, increased intensity of floods and droughts, and mangrove forests disappearing at an alarming rate (Cruz et al. 2007). When the prediction of the pace of the Himalayan glaciers melting was proved wrong, however, the report was rejected in its entirety by the Government of India for being too alarmist and "unscientific". This was later nicknamed "Himalaya-gate" (see Mathur 2001; *The Guardian* 2010) and initiated a policy move towards climate change scepticism, which for years had Indian politicians struggling to position themselves between the need to develop the national economy and the need to collaborate with an increasingly unified panel of global experts that required India to reduce emissions. As projections have become ever more dismal and the scientists more unanimous, overt public scepticism has, however, faded, and India has crafted a thorough policy of a transition to renewable energy – India's National Action Plan on Climate Change (N.A.P.C.C. 2010). During the past decade, the Indian state, I.P.C.C. reports and international attention has, despite certain scandals along the way, caused India's public to be more attuned to the current state of the Himalayas and the Shivalik Hills.

Although climate change is suspected to impact the already-stressed forest and water situation of India, with regards to agriculture, animal husbandry and health, the Indian Ministry of Environment seems to find the highest cause for concern in the Shivaliks. With the projected rise in temperature, wheat crops are expected to suffer a decrease in yield by 20% or more. Moisture deficits are also projected to lead to higher incidents of stem-rot, which will threaten the cultivation of the Indian mustard plant. The Haryana State Action Plan from 2011 warned that heat stress may lead to a decline in milk productivity amongst buffaloes and cows, that heat waves are predicted to cause an increase in human deaths from cardiovascular diseases, and that air pollution may lead to respiratory diseases. This will add to the already major outbreaks of infectious diseases transmitted by water (Haryana State Action Plan on Climate Change 2011:xxi). Raising "awareness" of environmental issues in a context of climate change has been the governmental focal point for over a decade, and the need for "awareness" in all sectors is explicitly stated (ibid.).

After the dramatic flood in Uttarakhand in 2013, described more fully in Chapter 6, the connection with global warming and the deteriorating Shivalik Hills environment became particularly salient. The flood was often portrayed in the news as a sign of what was to come, vaguely at first, but by my re-visit to the field by 2016, the environmental issues of the hills were fully encompassed in climate change rhetoric. For precautionary measures, N.B. Rao, Chairman of the Centre for Media Studies, was quoted in *Times of India*, saying that workshops on Climate Change Awareness would be held in all 12 mountain states "to raise awareness level among media and the related organisations who play key role in sensitising people about climate change". The need for this, he said, was that:

> Despite the fact that climate change has emerged as a huge issue world over, there is ambiguity about the issue among journalists and masses who need to understand it and espouse the cause with passion for [a] better tomorrow.
>
> (Rao 2016)

Sharing knowledge through awareness campaigns, as well as transferring technical solutions aiding human use of their environment, comprised a major part of the regional policy in the state of Haryana, whose northern borders include the hill village of Rani Mājri. In terms of which villages are suited for "sustainable development", Rani Mājri is a good candidate. The region in which the village is located is regarded by State Government measures as being socio-economically underdeveloped. Many villages here, such as Rani Mājri, still lack basic infrastructure and modern facilities, such as basic toilets, year-round access to drinking water and higher education. The last is especially pronounced amongst women as only 56% in the district are literate. There is still a high incidence of poverty and a seriously skewed sex ratio (837 females per 1,000 males) caused by female foeticide (Yadav et al. 2008:1; Sarv Shiksha Abhiyan 2015; Shivalik Development Agency 2017). This area also suffers severe environmental issues, such as the hills experiencing acute issues of erosion and deforestation. By all measures, the hillsides here are considered particularly ecologically fragile (Yadav et al. 2008; Haryana State Action Plan 2011; Chandigarh Administration 2015; Shivalik Development Agency 2017).

And so, various organisations, such as the United Nations (UN) and the European Union (EU); government authorities; non-governmental organisations (NGOs); and foundations have drafted plans and projects to draw attention to the state of these hills and the risks to their population in the face of climate change. Researchers map and register how in need of awareness these Shivalik Hills citizens are, while politicians and organisations arrange awareness campaigns so that inhabitants are empowered to conserve their immediate environment and learn why it is urgent that they do so *now*. This combination of "unaware" and "underdeveloped" people in an eco-sensitive area has made governmental and internationally financed Sustainable Development and conservation initiatives flourish in these hills. In Rani Mājri during my fieldwork in 2013, several actors were supposed to transmit "environmental awareness" and ensure "sustainable development". Most notable was the Indian Institute of Soil and Water Conservation (I.I.S.W.C.), operating under the Indian Council of Agricultural Research, the Haryana Department of Rural Development and the Haryana Forest Department as well as the State Government, which, by 2016, had implemented Eco-Clubs in 2,850 local village primary schools of the region.

Given the situation, educating people on the precarious balance that allows for life and prosperity on this planet seems like the right thing to do. It is important to plant trees, save ecological diversity by protecting areas from industrial development and teach people that their actions have larger consequences so that they act locally and think globally! As I settled in with the people of Rani Mājri, however, I realised that those ambitions were perhaps as naïve as they were Eurocentric. I quickly discovered that the villagers were never really confronted with climate change at all as they were thought to be too illiterate and uneducated to comprehend its complexity. The issue was, to echo I.I.S.W.C. scientists, a loss of translation from policy to practice. The "knowledge gap" between the scientific climate change idea, which all that policy and all those awareness campaigns intended to close, was part of a discourse that reinforced socio-economic differences between landholders and the landless, shaped environmental citizens as a people without knowledge and effectively muted an alternative perception of the *whys* and *hows* of climate change.

A scientifically social climate change

Climate change changes many things. One of them is how people perceive the relationship between humans and the world in which they dwell. This is nowhere so noticeable as in science itself. Scientists now argue for renaming the age we live in "the Anthropocene",[1] which denotes that the human effect on planet Earth is more decisive than geology. The term was popularised by Eugene Stoermer and Paul Crutzen (Chakrabarty 2009a; Haraway et al. 2016), and also denotes that humans have inadvertently become a major destructive force, likely to eradicate the life systems upon which we depend to survive (see Baer and Singer 2014; Haraway et al. 2016).

Increasingly, through the 1990s and up until 2020, unified groups of experts have vocalised quite strongly that the tempo of these changes is "very likely caused by human activity" (Crate and Nuttall 2009:10; Hijioka et al. 2014). In responding to the scientific fact that the globe is becoming warmer, that knowledge becomes immersed in a social world, with historical depth and political content. If humans contribute to the overproduction of greenhouse gases in the earth's atmosphere, then they need to adapt or mitigate those emissions by radically changing (a) the way we produce energy, (b) how much energy is used or (c) both. Mitigation is thus essentially an issue of how to regulate or change the production of what causes the emission of greenhouse gases, and adaptation is imperative to deal with the changes. Now, when climate change has become an issue of how and what we produce and consume, it follows that resource extraction and natural resource management must be adjusted to adapt or mitigate unwanted changes to our biosphere. The practice of natural resource management, here; the protection, ownership and allocation of the Earth's inanimate and animate riches, has however aggravated tensions and entered deals between people, groups, states and countries long before we made global theories on how we best should deal with them. These issues have been intrinsically socialised, culturally contextualised and ferociously debated in anthropology and later political ecology for decades. In the current situation of climate change-related issues, we see the same process, with researchers moving from proving the urgency of the issue, to "why" they happen and "how" to deal with it. In that transition, climate change too has become acknowledged by scholars as an intrinsically social and politically sensitive issue, merging into a patchwork of pre-existing ideologies and cosmologies on how to best thrive as a human in this world. What is also becoming increasingly apparent is that climate change is not just about what happens when there is too much carbon dioxide in the atmosphere. It can be argued that what people now, in fact, are agitated about is not climate change as a physical process per se but rather the *effects* of climate change on everything else. The concept of climate change in 2020 is a rich one, present in meeting rooms and political rhetoric, affecting patterns of mass consumption and carrying with it an uneasy juxtaposition of development and growth, preservation and equity. This means that, today, it is not just a rephrasing of what we in the 1990s talked about as "the greenhouse effect" or global warming; it has become something more than a scientific description of an extremely complex sociopolitical process.

This also revives doubts about the possibility of ever achieving the idea of Sustainable Development in its ecological, economic and social guises. Without resorting to dictatorship, an operative democracy will require that people still have the freedom to choose, preferably between the "right" thing in a long-term view and an unknown threat in an uncertain future. The urgency to ensure the continued existence of our contemporary worlds, give extra haste to how quick the "rest of the world" (the people of so-called "developing" nations and the currently industrialised countries) is expected to realise and adapt to the extent of the problem. Hence, a need for propagating environmental and climate change "awareness" from the global north to the global south again arise. It is generally accepted that awareness of both climate change and its effects on the environment is needed to mitigate the related issues, justifying political attention and funds being spent to make sure the diffusion of knowledge happens correctly. This is a prime example of what Adams (2009) calls "the fusion of environmental populism" (that people should be empowered through bottom-up approaches to development) and "ecological modernisation", where the optimal, technical and scientifically informed regulation of the state and market will create Sustainable Development (Adams 2009). This, I argue, is derived from how human-nature relations are approached in the Greco-Roman tradition, also referred to as "Western". I cannot possibly do justice to the full debate on how to define the nature-to-culture relationship here as these two concepts are inextricably tied to the development of Western science and philosophy, but, as Descola (2013) and Cronon (1996a,b) argue in their historical readers on the relationship between nature and culture in Western history, there is nothing natural (as in "given") about the concept of "nature", and neither is there [any]thing natural about the concept of wilderness" (Cronon 1996b:73, see also Dove 1992). During the 1980s–1990s, however, "nature" increasingly became a resource to be governed through being recast as the "environment", juxtaposed in an uneasy relationship with development in "sustainable development" (Adams 2009; Sachs 2010; Dryzeck 2013, see also Chapters 3 and 5).

Writing climate change

Arguing, as I do, that there are different conceptions of what nature or the environment constitutes does not mean that global warming and its potentially disruptive effect on terrestrial life is not *real*. But to argue that climate change also relates to human-environment relations in being a social, political and economic issue is a relatively recent addition to the field of science. Up until around 2010, social scientists interpreted the apparent public apathy towards scientific global warming data as a cognitive mismatch between the abstracted knowledge and the immediate perception of climate change. Aware of the dismal facts of climate science, scholars wrote lengthy pieces on why people seemed so indifferent to it, suggesting that, as a scientific fact, climate change was perhaps too abstract for people to agitate upon (Ingold 1995, 2000, 2010; Chakrabarty 2009a,b; Hulme 2009). Others suggested that people retreated into various states of denial (Norgaard 2011). In 2020, those *whys* and *hows* of climate change appear to spur people all over the world into action. Many people ferociously recycle their plastic, choose vegetarian or locally produced foods, attend school strikes for climate[2] or even spend their spare time "plogging"[3] in their local neighbourhoods. Others protest the idea that human emissions even matter to climate change and call it a left-wing, socialist propaganda machine. Researchers apply for grants to find correlations and explanations for the effects that education, nationality, age, gender or political ideology might have on how and why climate change continuously vexes, troubles and engages people into either action or inaction (e.g., Leiserowitz and Thaker 2012; Bertoldo et al. 2019). Simultaneously, politicians and investors line up to capitalise on and reinforce new trends, spurring more or less "climate friendly", "carbon neutral" and "green" enterprises. Climate change itself does not do these things. How people believe climate change to happen does.

Discourses on climate change, however, engulf us scientists too. Leichenko and O'Brien (2019) identify four discursive typologies within which Western society tends to view climate change, of which three might be discourses under which scientists working on climate change and society belong. The *biophysical discourse,* which addresses climate change as an environmental problem caused by excessive greenhouse emissions, speaks well to most natural scientists, perhaps; meanwhile, the *critical discourse,* which argues that climate change is a social problem caused by inequality and unsustainable development trajectories, speaks to social scientists and those who study the humanities. However, I argue, like Baer and Singer (2014:5), that, within anthropology, the discipline to which I belong, there are no clear approaches. Definite typologies are, thus, difficult to draw (Baer and Singer 2014:68). Baer and Singer suggest that anthropologists who work ethnographically on climate change issues do so through mainly four approaches. The first includes *cultural ecological approaches,* which predominantly concern themselves with risk and the capability of human social systems to adapt to potential new climate change scenarios. These studies might look at how resilient social networks can be and to what extent culture, or cultures, have the social mechanisms to cope. An example is the studies of anthropologist Marcela Vásquez-León (2009), who deals with climate change and changing Agricultural practices amongst Hispanic farmers in South-Eastern Arizona. Whilst Vásquez-León finds the farmers capable of adapting (to a certain degree, at least) to a changing climate, it is mostly through the efficacy of social resilience. As such, her study enriches our understanding of how aspects of sociality relate to how people or groups can adjust or mitigate to climate change, encompassing social problems and inequality with the biophysical world. The second approach is the *cultural interpretive or phenomenological approach,* which focusses on the construction of meaningfulness, and the perceptions and the interpretations of a changing environment. These ethnographies often explore the relation between "local" knowledge and climate phenomena (Baer and Singer 2014:67, 71), such as Ingold (2000, 2010) and Krause (2013). Within a phenomenological approach, climate change becomes perceptible to the human senses through changes in the weather or environment. All humans, especially perhaps Agriculturalists, tend to plan their activities because they expect certain kinds of weather from the season. To bring the abstractness of climate change "down to earth", so to speak, has been a call that many have responded to by looking at people's perceptions of weather and seasonality. Large parts of my own analysis draw upon this tradition, but, as Baer and Singer (2014) object, the phenomenological approach often tends to avoid addressing how perceptions of the environment and climate might relate to power. The latter is addressed ethnographically by the *critical approaches*, which are perhaps aligning with

the field of political ecology (Robbins 2019) and the critical discourse suggested by Leichenko and O'Brien (2019). Baer and Singer lastly introduce a somewhat all-encompassing *eclectic approach* (Baer and Singer 2014:5), under what Leichenko and O'Brien call the *integrative discourse*, where climate change is viewed as an environmental and social problem rooted in "particular beliefs and perceptions of human-environment relationships and humanity's place in the world" (Leichenko and O'Brien 2019:43). The eclectic approach is exemplified by Susan Crate's (2008) work on the Viliui Sakha of the Russian Siberia. She looks at how local perceptions of a changing climate (and the melting of the perma-frost) are expressed in narratives about the disappearance of the "Bull of Winter", a legendary creature that marks the advent of spring. Quite phenomenological in her approach to the Viliui Sakha's disappearing Bull of Winter, Crate, however, also addresses issues of climate adaptation and migration, and risk management, thus mixing political issues with cultural "sense making" in one piece. It is to this latter category of ethnographic research that I see my own work contributing.
....

A note on methodology

Every account of any place, history, people, and/or opinions is subjective to the researcher and his or her interaction with others in "the field". As an ethnographic fieldworker, one never escapes the social world one aspires to study and describe. Some do that job better than others, and I have been deeply inspired by works such as Gloria G. Raheja and Ann Gold's (1994) North Indian village studies on gender and kinship; Lila Abu-Lughod's (1999) study on gender, poetry and resistance from a Bedouin village in Egypt; Sarah Lamb's (2000) study on the aging Indian body in a village in East India's Bengal; and Anna Tsing's (2005) ethnography from the Meratus mountains of South-East Kalimantan, Indonesia. These authors have all produced thorough and complex ethnographies that have allowed me to reconfigure my own view of the world and the people with whom I share it. Acknowledging this, I thus owe the people of Rani Mājri and the readers of this book a brief introduction of myself, how I came to the information I did and what considerations I had when implementing it in my analysis.

Choosing a place for fieldwork

The idea of a suitable "place" for fieldwork was shaped by the early days of anthropology, when a male, fair-skinned, Western anthropologist did his fieldwork alone, in a remote location, with manuscripts that his research subjects would never read. The times, and the discipline, have changed a lot since then. Today, an anthropologist does not have to study in a remote location, nor must she be male or fair-skinned. The people we study increasingly read, react and collaborate with us to produce the analysis, and the idea

of the "local" is being continuously reinterpreted and redefined as encompassed in the "global" (see Hastrup 2013). These days, most anthropological fieldwork is, to a certain extent, "multi-sited" as people are increasingly
"multi-sited" too. Most people move about – either at will or out of necessity, over long distances or short – and through their movements they relate
to larger, more complex systems, often virtually as well as physically. To
trace local as well as global interconnections without losing ethnographic
depth is, thus, a challenge.

I have attempted to do so from a village for many reasons. First of all,
gathering data in one relatively small and nucleated village has its advan-
tages. One gets the benefit of "being known to everyone", which saves time in
letting people know what you are doing there. Based on previous fieldwork
experience from India (rural Rajasthan and Delhi), I also imagined a village
would be less chaotic – and even a safer and better place for my firstborn –
than the rather chaotic Indian cities. Having a rural, Hindi-speaking population, being a global climate and environmental hot-spot and having the
presence of larger cities with renowned hospitals in case of an emergency,
the area around Chandigarh emerged as a suitable site. In October 2012,
my husband, my seven-month old baby and I thus left Norway for Chandigarh, with a short stay in Delhi to arrange our plans. Chandigarh was
built to serve as the joint capital of the states of Punjab and Haryana in
the 1950s, and is well known for its cityscape being designed by the famous
Swiss-French architect Le Corbusier. From the hotels and large villas of the
more wealthy areas, or sectors, one can look directly up onto the hills of the
Shivalik, rising towards the Himalayan mountains. The hills would, from
that distance, appear lush and green, and the city attracts a great number
of tourists wishing to take in the view of the Shivalik Hills from hotels such
as "The Shivalik View" and "Mount-view Hotel". This seemed like a good
place to look for a suitable village in an Eco Sensitive Zone. After a few
weeks, I became impatient. How to make that first contact? Finally, in early
December 2012, a friendly hotel receptionist gave me the contact details of a
relative who worked with a team of scientists at the I.I.S.W.C. office in town
and who agreed to meet me. He promptly introduced me to his colleague,
Dr Swarn Lata Arya. Dr Arya is a renowned social scientist, responsible
for analysing and mapping the socioeconomic situation in those villages of
the Shivalik Hills that were eligible for the establishment of a state-funded
watershed management project. Rani Mājri was one of the villages she had
stayed in, and she had established good relations with the farmers there a
few years previously, when such a watershed management project was initiated there. This watershed consisted of a slightly elevated mountain ridge
covering five villages, with Rani Mājri sitting at the highest elevation. In
2012, the five-year watershed management project was nearly completed,
but the scientists who implemented it were still involved in some follow-up
activities in this village and intermittently visited it to oversee and partly
finance the restoration/improvement of a traditional water and irrigation

channels (*kuhl*). I was thus invited along for one of these visits in December, which gave me a chance to meet Prakash and his family. Prakash was a middle-aged farmer of the Rajput caste, who was just about to complete the construction of a new annex on his parents' joint household. It was due to be ready just before the wedding of his youngest brother in January, and, as such, Bhagwati and Bhupati, Prakash's elderly parents, invited us to live in that room with our little family. And so we did.

I am humble and grateful to the people of Rani Mājri for allowing me to outline their relationship with their surroundings and to expose their fears, dreams and ambitions. I have tried my best to take all ethical issues into consideration before, during and after my research was conducted. In those cases where the information felt "off the record", I have omitted it. In the cases where I was unsure if the person really understood the consequences of telling me this or that information, I have generalised it to shroud their identity. In general, however, I have related the information I gathered through interviews and participant observation with informed consent as the kind of information people would like to share and emphasise in meeting with me.

An ethnographic approach to climate change

The fieldwork I carried out in 2012–13, with a short follow-up visit in 2016, was a piece of qualitative fieldwork based upon participant observation, which really is a compilation of many methods (Atkinson and Hammersley 2010[2004]:31). It involved structured (formal) and semi-structured (informal) interviews, surveys and questionnaires; collections of life-stories; photography; and a lot of hanging around, trying to be of use and not a nuisance. In practice, this method of data collection requires that the researcher spend a lot of time with the same group of people, observing what they do and say, and at the same time engage in selected daily activities (Emerson et al. 1995:1). Participant observation, then, involves the dual practice of observing and engaging with people, which is a challenging exercise. As Norris (1993:126) points out, however, it is more useful to look at "participant-observation" as moving along a scale that "covers a continuum from complete participant to complete observer" and that varies with context and intent.

At times, I would do formal interviews of officials, members of NGOs and Governmental Organisations and other scientists and researchers working on either rural development and/or the environment in the region. I also utilised semi-structured interviews as a method of getting an impression of how environmental issues were dealt with and talked about in the public and urban sphere by policymakers. These more formal interviews were carried out face to face at times in the presence of other employees, like my interviews with the head of the Rural Department at its regional office, and at other times in semi-official contexts, as with NGO workers at their work site in Chandigarh or with government middle-school teachers on the school's premises. Sometimes, what began as formal interviews would become more informal with time and take place in more relaxed settings. This was the case with the interviews I conducted with the retired hydrologist, former government worker and now environmental NGO founder R.C. Gupta. These interviews were largely in English.

Social scientists doing qualitative fieldwork was a strange and unfamiliar concept in the village, however, and with time I got the feeling that the villagers forgot why I was there. Many would approach me as a friend or neighbour, sometimes forgetting my role as a researcher. I quickly realised that the local expectations of what a proper scientist did were based on the methods of a quantitative scientist – one that counted, measured and calculated data presented in neat graphs and tables. I would thus usually alternate between doing participant observation and doing questionnaires and informal/formal interviews. The latter method was useful both to provide information on landholding, power and status, but also served as a reminder to the villagers of my role as an "outsider" with a purpose to collect ethnographical data.

To conduct surveys whilst residing in the village was ethically one of the more complex methods I used. In the beginning, my skill in Hindi and *Pahaṛi* (the local dialect, also the name of the cultural identity that the villagers used to define themselves) was so low that I needed help translating answers. I hired Prakash's oldest daughter, who spoke a little English, and she would help me find Hindi alternatives for the broad dialect the elders spoke. On our first round in the village, I began with asking how often people used the forest, what they used it for and the likes. This worked well enough, and people readily answered. But in the second half of the questionnaire, I touched upon subjects that concerned the household economic situation, and the atmosphere quickly shifted from friendly chatter to muttering. Obviously, here, as in many other cultures, one is not supposed to discuss family finances publicly or with anyone but close kin, which explains, in part, the short and rather awkward ending of these interviews. For a fuller explanation however, I had to learn much more about life, statuses and relationships in the village. Only in retrospect did I realise that when I asked, "How many electrical appliances do you (or father-in-law/father) own?" and "How many animals do you (or your father-in-law/father) own?", I was asking this of people who own very little and who place great status and pride in the quantity of such commodities to demonstrate their wealth to an unknown, white, urban researcher. I exposed their status as "poor" in front of myself and to a landowner's daughter. The answers they provided were thus given to a member of a high caste and landowning household with a certain degree of influence in the village. Her caste and lineage status as a landowning Rajput, and my close association with them, could act, in a way, as a "gatekeeper" to knowledge (Berreman 2012). This was especially relevant in the structural and personal marginalisation experienced by the local scheduled caste (S.C.) population, for whom my initial questions must have borne a very strong resemblance to all the other surveys that had been done in the same region by the State Government, when the amount of "things owned" reported could have dire consequences.

Many of the families I interviewed were concerned that I would "uncover the truth" about the fridge or the television set they had received as part of a dowry. If the State Government census officers knew this, that extra appliance could tip the scale so that the household would fail to qualify for Haryana's "Below Poverty Line" (BPL) Index (see Chapter 2), risking their ration cards for food grains, for example. Luckily, long durée fieldwork provides the luxury of learning by doing, and I was able to change both my approach and the way I conducted my surveys and interviews along the way. By focussing more on participant observation and less on formal interviews, by gradually being able to ask questions without an assistant (except my toddler son, whose outgoing playfulness opened so many doors) and by repeatedly assuring my interviewees that I would report nothing to the district government, I helped people to relax and open up. Another implication of treading the border between friend and researcher like this is that one will likely observe so many "potent" situations for an ethnographic "case", but very often they are too personal, too compromising or too politically sensitive to be included in the analysis.

Opening and closing gates of information

Who you are, and through the people you associate with, you open and close gates of information in the field, often unwillingly and sometimes unknowingly. As the presumption of my relationship with the government faded, for example, I rather became more associated with the Rajput caste. In Rani Mājri, these caste relationships were central to defining whom I could spend time with. My visits to the S.C. hamlet was tolerated by the Rajputs among whom I lived, but I was told not to accept any food or drink in their house as "their food would make me ill" (see the Chapter 1 section on caste). Through that high-caste association, I thus lost access to participating in the daily lives of the S.C. population.

Additionally, I was a female researcher, married, with a child. Most of the material in my research was derived from conversations with women (mostly Rajput but also Lohar and S.C.) and practical engagement in women's work (see the Chapter 1 section on gender). This included most farm and household-related activities, except ploughing and certain rituals. To be a foreign, highly educated and Scandinavian married woman with a child in the village thus allowed me to participate in certain spheres of work and social life but regulated and restricted my participation in others. That meant that if there were public affairs to be handled, bills to be paid, documents to send or receive or court cases to settle, I was not always made aware of those, as men did those activities and seldom shared that information with the women of their homes.

To outline briefly how these gendered structures affect the rights and privileges of women in Rani Mājri would be useful at this point, both because they matter to the practicalities of fieldwork and because they matter to the information upon which I base my analysis. The women's self-categorisation particularly and the village population generally talking about themselves as "backward" and inferior did affect how I would go about "research". In the first couple of months, upon asking women the names of various animals, why they would use this tree instead of that or the political situation of the state, they would almost always claim to "know nothing". If I asked why some chose a stick from the neem tree over a toothbrush, why they would give me caraway for my stomach pains or why they added organic manure to a field every fourth year, the most common response was that I should ask the *paṛhe-likhe log*" – the educated people – or their husbands or teenagers who went to school, not the woman herself. It didn't occur to me at the time that they saw their kind of "knowing" as representing all that made them "backwards" in the eyes of the educated people, who knew the "correct" reasons for these things. Consequently, during the first few months of my fieldwork, women would defer to husbands so as to give the "Western researcher" a satisfactory answer. With time, however, and as we became better acquainted, the women became more willing to share their knowledge with me. Sometimes for my own good (so as to prevent me from making a fool of myself or them) or so that I would not inflict harm on my son or my husband unknowingly, but hopefully because I never saw them as "backward", and they knew that.

My marriage, and the presence of my husband in that respect, was a gate-opener in the field. In practice, I conformed to many of the same rules as a married woman returning to her natal village (*māykā*) for a holiday or visits. When she does, she is relieved from many of the chores and rules of etiquette that she must endure in her in-laws' home. At her *māykā,* she does not have to conform to veiling unless her husband or anyone senior is present, she does not lower her voice so humbly nor is she expected to work. This meant I could engage in discussions with men in the village without a chaperone, for example, without too many eyebrows raised, but it also gave me a different set of norms to fulfil, contradict or even resist (see Chapter 1).

Another definitive gate-opener in the field was my son. In the respect of being an object of public interest (fair-skinned, chubby babies with – at the time – blond hair are seldom encountered there), he was the starting point of many a conversation with strangers who wanted to hold, cuddle and talk to him. Motherhood also gave me something basic in common with other married women in the village. The associated emotions and challenges one experiences as a part of motherhood bypassed the occasionally large cultural differences we faced in dealing with each other, creating a levelled field upon which friendships and understanding were built. These spheres of intuitive recognition were important because no matter how much we discovered we had in common beyond being mothers or wives or daughters, there were initially so many differences, both real and imagined, that are set into play when a secular woman born and raised in a Scandinavian social democracy interacts with relatively poor, Hindu, Indian villagers. As a culturally Christian and Scandinavian female, I could never escape being a very *different* kind of woman, one indeterminable with respect to caste and creed, and inherently *Angrezi* – an "Englishwoman" (my Norwegian nationality did not affect this ascribed identity). I would therefore like to emphasise that I do not attempt to portray any part of the following ethnography as absolute or indisputable "fact" about an individual or an issue – it is to be read as situated perceptions and analysis only. I am also humbly aware of the trust bestowed in me to communicate these fragments of people's lives to others through the pages of this book. Any breaches of this trust are on my shoulders only.

Choice of words and how they flow

I went to study how perceptions of climate change as a physical process become caught up in daily lives in a North Indian village. But how to ethnographically approach such an abstract idea? It would therefore matter for the analysis not only whom I spoke to but *how* I spoke to them. This is why, at the outset, I decided not to discuss "climate change", "global warming" or "the environment" unless the person I talked to brought it up. That seldom happened, for two main reasons. First, it took time for me to expand on my basic Hindi with a mix of Pahari so that I could engage with ease in small-talk and slow-paced discussions. Second, these English phrases amongst non-English speakers served little purpose as they made almost no sense to those with only a few years of formal education. As an example, if I spoke to males about political and economic issues, weather patterns, soil quality, shrines, deities or farming practices, they would readily share their knowledge with me. But if I attempted to ask about the environment, even otherwise well-spoken fathers would call upon their sons or daughters – even the youngest of their schoolchildren – to answer the social scientist in their yard. "Environment" and "global warming" were new words for them, associated with school curricula.

I would at first attempt to explain that my research was about "nature". Referring to the passage on the conceptualisation of "nature" in Western science above, I was quickly reminded that this was not a viable option. "Nature" is sometimes translated as *prakrti* or *prakriti*, but is perhaps better translated as "source", and the concept is endowed with Hindu concepts of matter in its germinal state. As such, no direct translation of *nature* could carry the same connotation that it does in Western cultures. I did, however, note that in Rani Mājri, people would use *vātāvaraṇ*, (here *vatavaran* for readability) when talking about their immediate surroundings. This word literally means "wind covering", similar perhaps to "atmosphere", but it is also often used in the same way as weather (*mausam*) (Zoller 2016, personal communication). In this, it is perhaps more likened to the notion of "landscape" that – in line with Ingold's (2000) study – is not a totality to be observed, but a world in which we are immersed (Ingold 2000:207). I thus gave up English words entirely and rather asked about their seasons, how weather and monsoon patterns affected their crops, and how they regarded their surroundings. I also talked to them about their children, childhoods, husbands, concerns and hopes for the future. People answered with stories of progress, modernity, pollution, lack of care and global warming. These conversations revealed all the complexity of what "awareness" might be about and brought me closer to how climate change, as the complex process of the heating of our globe and enhancing the ongoing ecological crisis with it, is understood, interpreted and dealt with.

Notes

1 According to the British Geological Survey (B.G.S.), for example, humankind at the present lives in the Quaternary period, divided into the geological epoch of the Pleistocene (approximately 1.8 million years ago) and then sub-categorised into the Holocene (which encompass the last 11,700 years of the planet's history). The Holocene is characterised geologically in the extent of sediments deposits on land and sea, and it encompasses the rise of post-Stone Age humans (B.G.S. 2017).
2 School strikes for climate, alt. Fridays For Future, is a movement begun by the Swedish youth Greta Thunberg in 2018. The following year, students increasingly skipped school on Fridays to protest, demanding action from their political leaders. In 2019, millions of protesters marched in climate strikes in more than 163 countries (Encyclopaedia Britannica, 2020).
3 Plogging: from Swedish 'plocka och jogga', lit. picking up plastic waste as you walk or run.

Bibliography

Abu-Lughod, L.
1999   Veiled Sentiments: Honor and Poetry in a Bedouin Society. 2nd edition. Berkeley: University of California Press.
Adams, W.M.
2009   Green Development. Environment and Sustainability in a Developing World. 3rd edition. Oxon and New York: Routledge.
Agarwal, A. and Narain, S.
2000   Redressing Ecological Poverty Through Participatory Democracy: Case Studies from India. *PERI Working Paper Series* (36): 29.
Atkinson, P. and Hammersley, M.
2010   Feltmetodikk: Grunnlaget for feltarbeid og feltforsking. Oslo, Norway: Gyldendal Norsk Forlag.
Baer, H. and Singer, M.
2014   The Anthropology of Climate Change: An Integrated Critical Perspective. London and New York: Routledge, Earthscan.

Berreman, G. D.
2012   Behind Many Masks: Ethnography and Impression. In *Ethnographic Fieldwork: An Anthropological Reader*. C.G. Robben and J.A. Sluka, eds. Hoboken, NJ: Wiley-Blackwell.

Bertoldo, R., Mays, C., Böhm, G., Poortinga, W., Poumadère, M., Tvinnereim, E. and Pidgeon, N.
2019   Scientific Truth or Debate: On the Link Between Perceived Scientific Consensus and Belief in Anthropogenic Climate Change. *Public Understanding of Science* 28(7): 778–796.

Chakrabarty, D.
2009a   Dipesh Chakrabarty – Breaking the Wall of Two Cultures. Science and Humanities After Climate Change. In *Falling Walls Lectures*. http://www.fallingwalls.com/videos/Dipesh-Chakrabarty--1225, accessed October 30, 2015.
2009b   The Climate of History: Four Theses. *Critical Inquiry* 35(2): 197–222.

Chandigarh Administration
2015   Master Plan 2030. Department of Urban Planning, Chandigarh. http://chandigarh.gov.in/cmp_2031.htm, accessed May 14, 2020.

Chansel, L. and Piketty, T.
2019   Indian Income Inequality, 1922–2015: From British Raj to Billionaire Raj? *Review of Income and Wealth* 65(S1): S33–S62.

Crate, S.A.
2008   Gone the Bull of Winter? Grappling with the Cultural Implications of and Anthropology's Role(s) in Global Climate Change. *Current Anthropology* 49(4): 569–595.

Crate, S.A. and Nuttall, M.
2009   Anthropology and Climate Change: From Encounters to Actions. Walnut Creek, CA: Left Coast Press.

Cronon, W.
1996a   Introduction: In Search of Nature. In *Uncommon Ground: Rethinking the Human Place in Nature*. William Cronon, ed. Pp. 23–56. New York: W.W. Norton.
1996b   The Trouble with Wilderness; or, Getting Back to the Wrong Nature. In *Uncommon Ground: Rethinking the Human Place in Nature*. William Cronon, ed. Pp. 69–90. New York: W.W. Norton.

Cruz, R.V., Harasawa, H., Lal, M., et al.
2007   Asia. Climate Change 2007: Impacts, Adaptation and Vulnerability. Contributions to Working Group II to the *Fourth Assessment Report of the Intergovernmental Panel on Climate Change*. M.L. Parry, O.F. Canziani, J.P. Palutikof, P.J. van Der Linden and C.E. Hanson, eds. https://www.ipcc.ch/publications_and_data/ar4/wg2/en/ch10.html, accessed September 14, 2015.

Descola, P.
2013   Beyond Nature and Culture. London: The University of Chicago Press, Ltd. U.S.A.

Dove, M.R.
1992   The Dialectical History of "Jungle" in Pakistan: An Examination of the Relationship Between Nature and Culture. *Journal of Anthropological Research* 48(3): 231–253.

Dryzeck, J.
2013   Introduction. In *Politics of the Earth – Environmental Discourse*. J. Dryzeck, ed. Pp. 3–25. 3rd edition. Oxford, UK: Oxford University Press.

Dubash, N.K.

2013   The Politics of Climate Change in India: Narratives of Equity and Cobene-
fits. *WIREs Clim Change*, 4: 191–201. doi:10.1002/wcc.210
Emerson, R.M., Fretz, R.I. and Shaw, L.L.
1995   Writing Ethnographic Fieldnotes. Chicago, IL: University of Chicago
Press.
Encyclopaedia Britannica
2020   Greta Thunberg. https://www.britannica.com/biography/Greta-Thunberg#
ref1276549, accessed June 6, 2020.
Government of India
2011   *Census 2011 India*. http://www.census2011.co.in/, accessed April 4, 2017.
Gupta, D.
2005   Caste and Politics: Identity over System. *Annual Review of Anthropology* 34:
409–427.
Haraway, D., Ishikawa, N. and Gilbert, S.F.
2016   Anthropologists Are Talking – About the Anthropocene. *Ethnos* 81(3):
535–564.
Haryana State Action Plan on Climate Change
2011   Ministry of Environment, Forest and Climate Change, Government of India. http://www.moef.nic.in/sites/default/files/sapcc/Haryana.pdf, accessed May 11,
2017.
Hastrup, K.
2013   Scales of Attention in Fieldwork: Global Connections and Local Concerns
in the Arctic. *Ethnography* 14(2): 145–164.
Hijioka, Y., Lin, E., Pereira, J.J., et al.
2014   Asia. In Climate Change 2014: Impacts, Adaptation, and Vulnerability.
Part B: Regional Aspects. Contribution of Working Group II to the *Fifth Assessment Report of the Intergovernmental Panel on Climate Change*. V.R. Barros,
C.B. Field, D.J. Dokken, M.D. Mastrandrea, K.J. Mach, T.E. Bilir, M. Chat-
terjee, K.L. Ebi, Y.O. Estrada, R.C. Genova, B. Girma, E.S. Kissel, A.N. Levy,
S. MacCracken, P.R. Mastrandrea, and L.L. White, eds. Pp. 1327–1370. United
Kingdom and New York: Cambridge University Press.
Himalayan Climate and Water Atlas
2015   Impact of Climate Change on Water Resources in Five of Asia's Major
River Basins. *ICIMOD, GRID-Arendal, CICERO*. https://www.grida.no/publica-
tions/69, accessed April 11, 2017.
Hulme, M.
2009   Why We Disagree about Climate Change: Understanding Controversy, Inaction and Opportunity. 4th edition. Cambridge, UK and New York: Cambridge
University Press.
Indian Brand Equity Foundation (I.B.E.F.)
2020   About Indian Economy Growth Rate & Statistics. https://www.ibef.org/
economy/indian-economy-overview, accessed July 3, 2020.
Ingold, T.
1995   Globes and Spheres: The Topology of Environmentalism. In *Environmentalism The View from Anthropology*. Kay Milton, ed. Pp. 31–42. London and New
York: Routledge.
2000   The Perception of the Environment: Essays on Livelihood, Dwelling and
Skill. Reissue. London and New York: Routledge.
2010   Footprints through the Weather-World: Walking, Breathing, Knowing.
Journal of the Royal Anthropological Institute 16: S121–S139.

2011   Being Alive: Essays on Movement, Knowledge and Description. London and New York: Routledge.

I.P.C.C.

2019   Summary for Policymakers. In *IPCC Special Report on the Ocean and Cryosphere in a Changing Climate*. H.-O. Pörtner, D.C. Roberts, V. Masson-Delmotte, P. Zhai, M. Tignor, E. Poloczanska, K. Mintenbeck, M. Nicolai, A. Okem, J. Petzold, B. Rama, N. Weyer, eds. Pp. 1–35, https://www.ipcc.ch/site/assets/uploads/sites/3/2019/11/03_SROCC_SPM_FINAL.pdf, accessed August 23, 2019.

Krause, F.

2013   Seasons as Rhythms on the Kemi River in Finnish Lapland. *Ethnos* 78(1): 23–46.

Lamb, S.

2000   White Saris and Sweet Mangoes: Aging, Gender and Body in North India. Berkeley and Los Angeles: University of California Press.

Leichenko, R. and O'Brien, K.

2019   Climate and Society. Transforming the Future. Cambridge, UK: Polity Press.

Leiserowitz, A. and Thaker, J.

2012   Climate Change in the Indian Mind. Yale Project on Climate Change Communication, Yale University.

Mathur, N.

2001   Myth, Image and Ecology. *Indian Anthropologist* 31(1): 19–28.

Met Office

2017   What Is Climate Change? United Kingdom National Weather Service. Met Office. http://www.metoffice.gov.uk/climate-guide/climate-change, accessed May 12, 2017.

Ministry of Agriculture & Farmers Welfare

2019   Categorisation of Farmers. *Government of India*. https://pib.gov.in/newsite/PrintRelease.aspx?relid=188051, accessed June 2, 2019.

Moore, H.

2014   Recorded Lecture with Henrietta L. Moore at the University of Oslo. https://www.sv.uio.no/sai/om/aktuelt/sai-50/opptakmoore.html, accessed December 16, 2015.

N.A.P.P.C.

2010   Department of Science & Technology, *Government of India*. National Mission on Strategic Knowledge for Climate Change. National Action Plan on Climate Change. New Delhi. https://dst.gov.in/sites/default/files/NMSKCC_mission%20document%201.pdf, accessed June 2020.

N.A.S.A.

2015   What's the Difference Between Weather and Climate? NASA. http://www.nasa.gov/mission_pages/noaa-n/climate/climate_weather.html, accessed January 12, 2016.

2017   Global Surface Temperature | NASA Global Climate Change. Climate Change: Vital Signs of the Planet. https://climate.nasa.gov/vital-signs/global-temperature, accessed February 20, 2017.

Norgaard, K.M.

2011   Living in Denial: Climate Change, Emotions, and Everyday Life. Cambridge, MA and London, England: MIT Press.

Norris, C.

1993   Some Ethical Considerations of Field-Work with the Police. In *Interpreting the Field: Accounts of Ethnography*. 1st edition. Timothy May and Dick Hobbs, eds. Pp. 123–141. Oxford, UK and New York: Clarendon Press.

Raheja, G.G. and Gold, A.
1994   Listen to the Heron's Words: Reimagining Gender and Kinship in North India. Berkeley, CA: University of California Press.

Rao, N.B.
2016   Interviewed in Times of India. Uttarakhand Facing Impacts of Climate Change. *Times of India*. http://timesofindia.indiatimes.com/city/dehradun/Uttarakhand-facing-impacts-of-climate-change/articleshow/55229828.cms, accessed March 10, 2017.

Robbins, P.
2019   Political Ecology: A Critical Introduction. 3rd edition. Chichester: Wiley-Blackwell.

Rudiak-Gould, P.
2012   Promiscuous Corroboration and Climate Change Translation: A Case Study from the Marshall Islands. *Global Environmental Change* 22(1): 46–54.
2013   Climate Change and Tradition in a Small Island State: The Rising Tide. Routledge Studies in Anthropology, 13. New York: Routledge, Taylor & Francis Group.

Sachs, W.
2010   The Development Dictionary: A Guide to Knowledge as Power. 2nd edition. First Published in 1992.Great Britain: Zed Books.

Sarv Shiksha Abhiyan
2015   HSSPP | SSA | RMSA | Haryana. http://www.hsspp.in/, accessed September 25, 2015.

Shivalik Development Agency
2017   Demographic Profile. *Government of Haryana*. http://www.sda.gov.in/Page.aspx?n=137, accessed April 4, 2017.

Shorrocks, A., Davies, J.B., Lluberas, R. and Koutsoukis, A.
2016   Global Wealth Report by Credit Suisse Institute. https://www.credit-suisse.com/about-us-news/en/articles/news-and-expertise/the-global-wealth-report-2016-201611.html, accessed June 20, 2020.

Swaminathan, S., et al.
2019   The Burden of Child and Maternal Malnutrition and Trends in Its Indicators in the States of India: The Global Burden of Disease Study 1990–2017. India State-Level Disease Burden Initiative Malnutrition Collaborators. *Lancet Child Adolescent Health* 3: 855–870. https://doi.org/10.1016/S2352-4642(19)30273-1, accessed May 18, 2020.

The British Geological Survey (B.G.S.)
2017   Palaeogene to Quaternary. *Discovering Geology*. http://www.bgs.ac.uk/discoveringGeology/time/timechart/phanerozoic/cenozoic.html, accessed May 11, 2017.

The Guardian
2010   IPCC Officials Admit Mistake over Melting Himalayan Glaciers. Carrington, D., and Wylie, D. *The Guardian*, January 20. http://www.theguardian.com/environment/2010/jan/20/ipcc-himalayan-glaciers-mistake, accessed September 19, 2017.

Tsing, A.L.

2005   Friction: An Ethnography of Global Connection. Princeton, NJ and Oxford: Princeton University Press.

Tuttle, R.H.
2017   Human Evolution. *Encyclopaedia Britannica Online Edition*. https://www. britannica.com/science/human-evolution, accessed May 27, 2017.

UNICEF
2019   The State of the World's Children 2019. Children, Food and Nutrition: Growing Well in a Changing World. New York: UNICEF.

U.S. Environmental Protection Agency
2017   Climate Change Indicators: Weather and Climate. Reports and Assessments. https://www.epa.gov/climate-indicators/weather-climate, accessed May 12, 2017.

Vásquez-León, M.
2009   Hispanic Farmers and Farmworkers: Social Networks, Institutional Exclusion, and Climate Vulnerability in Southeastern Arisona. *American Anthropologist* 111(3): 289–301.

Wish, V.
2010   Allianz Knowledge: Fact Sheets on India (a) India: Climate Fact Sheet, India: (B) Climate Impact on Environment and Society. *Alliance Knowledge*. http:// knowledge.allianz.com/?155, accessed November 2, 2011.

World Bank
2015   The World Bank: India. Country Dashboard India. http://databank.worldbank. org/data/Views/Reports/ReportWidgetCustom.aspx?Report_Name=Country_ chart1_June4&Id=5ee1de8357&tb=y&dd=n&pr=n&export=y&xlbl=y&ylbl=y&legend=y&wd=430&ht=380&isportal=y&inf=n&exptypes=Excel&country=IND&series=SI.POV.NOP1,SI.POV.DDAY, accessed September 14, 2015.
2019   The World Bank: India: At a Glance. https://www.worldbank.org/en/country/india/overview, accessed June 20, 2020.

World Wildlife Fund
2017   Himalayas. Places. World Wildlife Fund. http://www.worldwildlife.org/ places/eastern-himalayas, accessed February 22, 2017.

Yadav, R.P., Singh, P., Arya, S.L., Bhatt, V.K. and Sharma, P.
2008   Detailed Project Report: For Implementation under NWDPRA Scheme. *Central Soil & Water Conservation Research and Training Institute, Research Centre, Chandigarh*. Ministry of Agriculture, Govt. of India, New Delhi.

Zoller, C.P.
2016   Personal Communication.

1   Climate change expressions

Barsāt

The winds that carry with them the warm, moist air from the southwestern Indian Ocean are eagerly awaited in the village of Rani Mājri. The monsoon season of Barsāt *usually makes its appearance with the onset of the month of* Asāṛh *and lasts through the month of* Sāwan, *which corresponds to mid-June until mid-August in the Gregorian calendar.*

By mid-June, clouds had drifted tauntingly towards the Himalayan mountains for days, without shedding a drop of rain. The schoolteachers from city did not think the rain would appear any time soon, the news had reported possible onset dates later in June. Amongst farmers, however, there was talk that the rains were surely approaching. Bhagwati, Prakash's aged mother, had said that a circle around the moon was a sign of rain to come the previous evening. The increased intensity of heat over the last few days had also made several women remark that the rains would be over us soon.

The next morning, someone in the village had decided there were enough indications and initiated preparations for the first rains. This brought hectic activity to the whole of Rani Mājri and demanded much from everyone able to work. But would rain fall?

It did.

The first few days, just a little fell, a teaser of what was to come. Knowing the right time for tilling and sowing to commence is crucial. In between the first showers, the soil stays moist for a very limited span. The longer heat and dry spells in between them bring the soil back to the dry and cracked surface of summer, impossible for any oxen to plough or human to dig. During this period, families work the fields from sunrise to sunset, immersed in work and sweat. The men of the landholding houses, the majority of Rani Mājri's central village, plough and till their terraced land with a wooden plough drawn behind two bulls.

The women, and the children after school, carry two-year-old composted manure, now in the state of fertile soil, in baskets on their heads out on the fields. Emptying the baskets down on the newly tilled land, they mould the warm and airy manure into the soil by hand. There is far from enough manure to cover all the plots, however, so even the largest households will have to prioritise which fields will receive the organic fertiliser. In Rani Mājri, these fields are those irrigated by the kuhl, *the small irrigation system that covers the land directly beneath the village. On a four-year rotational basis, the farmers always prioritise enriching the fields dedicated to that year's ginger crop. The ginger, when dried into the marketable product of* soṅṭh, *is the main source of income for the largest landowners.*

The largest landowners have bought the roots from Shimla through a joint effort (their home-grown ginger rots in the summer heat), and they have spent the last weeks before the rains indoors in the heat, preparing blue crates of ginger-roots, which are carefully sorted into piles depending on share. Entire households attend the planting, some digging ditches and troughs for water-flow, others placing roots in a systematic pattern with other vegetables (turmeric roots and colocasia variants of the taro plant, as well as coriander). The careful pattern is intended to provide just the right amounts of sun and shade, water and nutrients to the ginger. The roots, all laid out in a pattern, are then pressed into the earth with a pry-bar. Finally, rice-grass from the previous season and leaves from lopped trees are put on top, covering every little corner, to protect the crop from the sun that peers down between these early batches of rain.

After the ginger is sown, there are other fields to tend to, those uphill or lying beyond the controlled irrigation of the kuhl *system, like the southern fields that, during the winter season, lie fallow. During the monsoon season, these fields usually receive enough rain to support a decent yield of pearl millet or maize, broadcasted on the ground and then ploughed into the soil.*

The physical strain of the labour is exhausting and time-consuming. In the glaring sunlight and high temperatures, sweat pours down and disturbs your vision. In the short breaks under a shadow of a lonely tree on the meticulously worked land, the high-caste landowners gather in the middle, with the landless and low caste daily labourers assisting them, sharing in the conversation from the sunny periphery. It is late night before the families return to their respective households for dinner. After a long day's work, people wait a little for the sweat to dry before taking a highly irregular evening bath (normally, water is saved for the morning routine). Stomachs growling, people sit in darkness on roofs, stretching to reach the faint winds, hoping to cool sore muscles, whilst the women of the house prepare the evening food. Rice and lentils, sour milk to drink – again. Food is unvaried, and fresh vegetables have not been available for weeks due to the hot and dry season of summer behind them.

Days pass, with only taunting clouds drifting above the village. The days that passed in between these first monsoon showers are hot and humid, with temperatures circling 40° C. The halting fans caused by recurrent electricity outages in the middle of the night leave a dark, warm silence, mitigated only by the village's altitude, which makes the night a few degrees cooler than in the cities of the plains. Then, finally, by mid-June, the intense heat is broken off by intense and heavy rain. The monsoon proper has arrived, and the month of Sāwan has begun. With that, a ritual was held for Khwaja Pīr, a deity of rain and water with significant powers. He is, when attended to, a benevolent protector for the village during the monsoon, and his benevolence is especially

needed now. So is the goodwill from other deities, and men and women –
unmarried and married – choose to work empty-bellied from sunrise to sunset
on Mondays throughout Sāwan *for Shiva and Krishna.*

After a few days, the air is so thick with humidity and the surroundings so
potent with growth, one feels like it is almost bursting. The fodder gets wet. The
laundry gets wet. The floors become muddy. After a few days of intense show-
ers, everything crawls with life – there are bugs everywhere, in the flour, in the
bed, in the bathrooms, in the air. Frogs croak, flies hum, mosquitoes buzz, but
the birds are quieter than ever. The absence of wind and the high temperatures
in between the rains make the sweat pour; itching heat rashes are accompanied
by frequent diarrhoea and vomiting, believed to come from working in the sun
or at times from drinking the monsoon water from the kuhl. *The rainwater's*
movement through the ground is known to bring bacteria from the excrement
of humans and animals further uphill, who defecate in the forest above. Also,
the kuhl *water is so interspersed with silt that, although in abundance, it would*
be difficult to drink. It was worse before, however: my old neighbour illustrated
that she would have to filter the water for sand and debris through her teeth
(which she was no longer in possession of). And the season's extreme humidity
coupled with extreme heat, fasting and waterborne diseases does certainly af-
fect people's moods. In Rani Mājri, girls who are in their first year of marriage
now leave their new homes for the entire month, as Sāwan *is a time when, in*
the first year of marriage, mother-in-law and daughter-in-law should not meet.
*Anger (*gussa*) is thought to easily arise, unnecessarily straining the new and*
important relationship.

The monsoon rain falls far from constantly but is marked by dramatic shifts
and turns. So are work and travel. The increasingly heavy monsoon showers
create distortions in communication, bordering havoc in the cities (The Trib-
une 2013). The streets overflow with rainwater, causing plugged sewers to brim,
tarmac to collapse and trees to topple over in the strong wind. Electricity lines
snap, traffic gets blocked. In the village too, work halts when heavy clouds
release torrential rainfall, making water fall as if from buckets. Strong winds
carry cool air and thunderclaps that make people hurry home or to shelter.
There are frequent landslides too, smaller and larger, as porous soil gets heavy
with lashes of rain. The showers are never that long, there is time to sit, to wait,
to listen. When the rain halts again, people resume their tasks, attuned to the
rhythm of the season.

As Sāwan *comes to an end, the rain does too. Instead of dramatically falling in*
buckets, it now drizzles a little every day. After almost two months of humidity,
the heat begins to let go, making the air feel almost pleasant, and the surround-
ings have become an explosion of green. There is even green where there should
not be any, like on the walls, rooftops and telephone lines and electricity cables.
The maize has grown from light to dark green, the chili plants are lush – the
landscape is barely recognisable from its deserted state only a few months ago.
In the late evenings, children flock to the mango-grove between the Scheduled
Caste hamlet and Khot, with plastic bags. Nimble, they climb up the majestic
trees, shaking the mangoes down to the youngest, who are left on the ground as a

gathering crew, trying to outsmart the sullen cows who feast on the fallen fruits. Ripe mangoes are indulged in on the spot; unripe mangoes are carried home and used for the local chutney, *intensely sweet and salty at the same time.*

This is also the right time for planting the final monsoon crop, the rice grass (dhān). *Once again whole families gather on the fields. The women fold their trousers* (shalwar) *to their knees, tie their tunics* (qamīz) *above the hips into a knot and tie the shawl (*chunni) *around their heads. The sandals are kicked off, their feet make squelchy sounds in the knee-deep water, swapping place with water and mud. Plop-plop: the rice grass goes into the wet soil below the water surface – not too deep, or it would not grow. Sposh-splosh – not too shallow, or it will not root. The light breeze adds a playful tint to the labour. Periods of quiet concentration are broken with periods of cheerful songs (*gīt). *During a tea-break, an "uncle" makes a tiny "windmill" from a stick and a mango leaf for a small boy, who whizzes about with it, making it go around and around. A few boys attempt to use the* kuhl *as a waterslide. Finally, there is rest in sight, and everyone gathers for dinner. These days the vegetable dishes are enriched somewhat with the ripening bottle gourd and freshly made mango chutney. Autumn is arriving.*

....

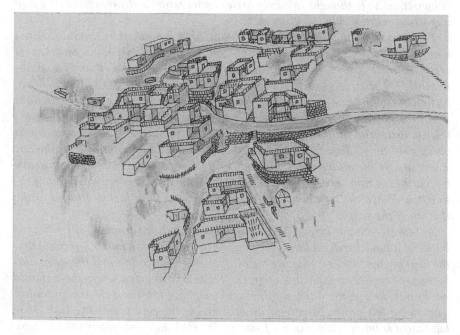

Figure 1.1 Drawing of the village.
Source: Author.
A hand-drawn map of the village indicating the size and approximate location of houses.

In Rani Mājri, on the outskirts of the alluvial plains of northern Haryana, subsistence farmers, shopkeepers, auto-rickshaw drivers, factory workers, sweepers and day-to-day labourers, and here, they dwell in a neat collection of houses. The village settlement is small, even in a hill village context, with just about 50 joint and nuclear households. These are organised roughly according to caste patterns, of which there are two in the main village, and one in a separate settlement (I will return briefly to caste and what it entails later on). Most households in the central village are small landholders, utilizing less than 1 hectare (800m^2 or 1 *bighe*, the local standard for land measurement). Some were landless, working other families land, and a few landholders occupied 2 hectares of land or above, which classified as a small landholder according to the Indian Agricultural census (Ministry of Agriculture & Farmers' Welfare 2019), but a large landholder in the local context. In this village, landholders and landless, farmers and factory workers alike, shared one thing. A lifestyle where their well-being hinged upon the environment being rather predicable and benevolent, and the large landowners' crops to yield bounty. As such, the daily life of Rani Mājri revolved around the Agricultural year, as it had done for as long as anyone could remember. Survival had never been effortless. The steep slopes of the terrain, the sandy quality of the soil and the amount of arable land has for centuries made it difficult to meet the demands of subsistence here. In these hills, as in the rest of India, the monsoon is an important ally. So decisive is the monsoon's effect on the national economy that a former finance minister and president of India, Pranab Mukherjee, allegedly called the Indian monsoon the true finance minister of the country (Narain 2014; Shira 2015). In the village of Rani Mājri too, the rain falling at the end of the hot and dry summer season provides relief from the intense summer heat, kickstarting the local Agricultural production, power generation and construction work. Since the early 21st century, however, the strength and pace of the monsoon had been more erratic than remembered by the living generation, and this has intensified a problem that has been an issue in these hills as long as anyone can remember: soil erosion. Every year, I.I.S.W.C. reports state, enormous masses of silt (150–650 tons per hectare per year) are being moved through these hills with water from rain and melting glaciers from the Himalayan mountains (Yadav et al. 2008). The movement cause landslides and soil depletion, as silt and nutrients of the soil move with the water in rivers and seasonal streams, down the deforested slopes of the Shivalik Hills, and are eventually carried on to the plains and into the sea. If the current projections of climate change effects in the Shivalik Hills are correct, the hot season of summer will be hotter, the dry season drier, and the monsoon rain will be more intense and falling at other times than expected. With unpredictable skies above their heads and the ground shifting below their feet, the dwellers of Rani Mājri has to adjust quickly, not only to a changing environment and a changing climate; but to a changing society too.

Life and how to go about it have changed more for the farmers here in one generation than is imaginable. In one generation, Rani Mājri went from no

health or education services, no road connection, no electricity and no landline telephones to become subjects for state development policies, schemes
and projects that has provided their children with public and private education services; public health services; electricity; water-grid access; televised
entertainment; accessible markets and mobile and satellite network access.
As people are vaccinated, educated and emancipated – from diphtheria, po-
lio, and smallpox, from extreme poverty, repressive norms and oppressive
traditions at a remarkable pace – they must also adjust to a globalized, liber-
alized market economy. With rapid shifts between top-down to bottom-up
rural development strategies and the increased need for productivity and
efficiency in a global trade market, combined with rapid shifts in prices for
produce caused by India's relatively recent economic liberalisation policies,
a good life in contemporary Rani Mājri requires just the right amount of
seasonal stability, knowledge, skill, time and money.

For most families here, that combination requires making tough decisions. If wild boar and deer destroy the yield, the farmer must invest in fences
instead of education for his or her children. If one season fails due to adding
too little fertiliser or not weeding of fencing meticulously enough, tragedy
can be avoided the next season, but always at a cost. A daughter's wedding.
A sick buffalo. Investing in more land is seldom an option, so whatever land
there is, it must be productive. As some land is reserved for forests, other for
industry and economic zones, farmers have little land to expand onto, and as
families grow, land become divided between sons, the next generation with
ever smaller plots than the preceding one. With less land to grow, there is less
yield, but not significantly more time to seek employment outside the Agricultural sector. As both men and women crop, weed, and fertilise the fields
by hand, and men till their land using oxen, the time spent on farming to
meet the demands of the household is significant. To meet the need for cash,
younger men increasingly depart agriculture as main source of income, and
seek employment in the city or in the factories. As their wives are left tend to
whatever land they have left, these young men commute to low-skilled factory employment, a journey easily traversed from the village in a few hours.
In these nearby industrial zones, they can apply for manual and/or conveyor-
belt work, spending day and night moving and packing soap and washing
detergents, motor-engine parts, etc. With a surplus of willing and able workers, the men were regularly fired (or re-hired), in flux with the market, the
demand for labour at the farm in the village, or other family or individual
issues. The precarious labour conditions of the factories made many aspire
for more steady, white collar occupations in the public sector, but these options were not for the majority. There were notable differences in caste and
status, which made that process of adjustment easier for some, than others.

Social principles of differentiation in Rani Mājri

Sharing the hearth with a joint family of two grandparents, their three sons
and their wives and children for six months, we awaited the monsoon's

arrival in 2013 with a shared sense of tense anticipation. The family with whom we resided in 2013 were subsistence Agriculturalists with a relatively small landholding, and, like most families in these hills, were dependent on a beneficial rainy season to meet their needs. A common concern for the arrival of the monsoon rain aside, it was the social, economic and political differences between the households here that enabled some families to adjust and mediate those very changes, whilst disabling others. These differences between individuals and families of Rani Mājri means that there are no "coherent hill village" to develop, but a society built up and segregated through friendships, alliances, and disagreement, in relationships often aligned and reinforced by social structures based ascribed and avowed identities – of gender, caste and class. This "multiplicity" of social principles of differentiation that constitute the complex fabric of relations between people, as well as between people and their deities, are often overlooked when crafting policy plans for climate change or Sustainable Development. This unawareness of local social structures often creates unintended, even undesirable consequences for such developmental initiatives as described in this book, and as such, these social relations and their structure are described below.

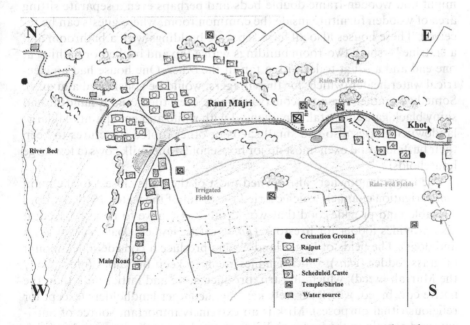

Figure 1.2 Approximate location of temples, fields, water sources and settlements in Rani Mājri.

Source: Author.

Map showing the approximate location of temples, fields, water sources and settlements in Rani Mājri.

Class and landholding in Rani Mājri

At first glance, the houses in the small main village of Rani Mājri look rather like each other. Most of them are painted in shades of a dusty yellow, blushing peach or faded turquoise, and many are small and built wall-to-wall. A closer look reveals that they are not so similar at all, and that there are stark differences in wealth and opportunities for social mobilisation This inequality explicitly relates to water access, which again relates to caste identity and landownership.

The large houses, as well as their landholdings, clearly follow the trajectory of the *kuhl*. At the highest elevation, and skirting the *kuhl's* trajectory from the hills, are Rajput landholder houses but for one single Lohar household. Mid-village, where the water submerges into the intricate irrigation system at a lower, south-western point, the Lohar-dominated settlement begins. These houses are constructed along the paved road, which slowly descends downhill towards other, equally small villages.

The houses that lie along this trajectory of the *kuhl* are relatively large, often two-storeys with anywhere between three and ten rooms (in the largest house). They are built back-to-back, expanded here and there, making it hard for a first-time visitor to separate one house from the next. Most houses have indoor kitchens for their hearth (*chūlhā*), gas-top stoves, and walls and floors made *pakkā* of cement and bricks. Inside the bedrooms one might find wooden-frame double beds and perhaps even a separate sitting area of wooden furniture inside the common room, where guests can be received. These houses also all have separate buildings with a bathroom and a "latrine" – small two-room buildings with a tap and bucket for bathing at one end and a squat toilet with a tap in the other. One house had an electrical water-heater (which, for the villagers, would be considered a luxury). Some have rather spacious courtyards, one even tiled, and a handful had shiny black gates in metal, ornamented with white and gold paint. All large landowner houses also had small sheds or other roofed structures to keep their buffaloes and oxen (most major households) and milk cows (a few large households).

The larger landowners also owned most of the plots closest to the main *kuhl* irrigation channel, which carried the brunt of the water. Not only does irrigable land provide food that would otherwise have to be purchased; it also provides fodder for livestock (*paśu*: cattle, horses, goats, sheep, asses and dogs). The fields of larger landowners produce a considerable amount of grass fodder (*ghas*), which enables them to keep buffaloes (most likely the Murrah breed), which, in turn, produce more and fattier milk than the native cow breed, which is rarely kept by the larger landholders (except for religious/ritual purposes). Milk is an extremely important source of nutrients and protein in the farmers' diets – not only does the sugar and caffeine in the milky tea provide them the energy to go full working days without a meal, but milk is also used in cooking, in vegetable soups, to make curd, in sour-milk drinks for the heat and often in the cooking cheese *paneer* (home-made paneer was not made in this village, however). But buffaloes eat a lot. For optimal milk production, buffaloes and bullocks are stall-fed to ensure an average daily diet of approximately 20–22 kg of green fodder and 5–7 kg of dry fodder (wheat straw, dried maize, sorghum and millet herbage) per day. The amount required for an adult, lactating buffalo is equal to the amount of fodder that would feed 7 adult goats (Arya et al. 1994:447). With access to self-sustenance excess fodder, large landowners could also keep bullocks (a humped cattle breed, most likely a Zebu species) – a necessity for tilling the land since no tractors could work them because of the terraced inclination and the way the plots are separated by stone fences. (A tractor was usually collectively rented for threshing and parked at the side of the road.) Bullocks would also require supplement nutrition in the month preceding ploughing, which had to be produced on-site or bought in stores. Except for milk production and tilling, keeping cattle was also important for the production of manure, which was used as organic fertiliser to make the irrigated lands more fertile. The amount of manure produced by the cattle is, in fact, quite significant. The Food and Agriculture Organisation of a United Nations study from Nepal showed that the amount of dung produced from cattle and buffalo per animal per day was "about 10 kg and 12 kg respectively giving a total of 3.6 tons and 4.3 tons of dung per animal per annum" (Joshi 2017). In addition, a landowner of reasonable size could trade surplus crops or even dedicate some fields for pure cash-cropping as a source of supplementary income. Many of the largest landowners collectively grew and sold ginger – dried and sold as *sonth* – and benefited from individual sales of surplus wheat and tomatoes. Rice and wheat would also be used as payment in kind for tractor and transport rentals or for services performed by scheduled caste people (S.Cs) and marginal Rajput villagers during harvest.

Although there was little to no possibility of expanding the amount of land under irrigation, there were opportunities to expand farming by "leasing" adjoining village land or by joining together when selling and buying ginger (which the village landowners did in unison). In this way, landholding families across castes and village borders could be in a reciprocal relationship with one another. For example, the household of Prakash and his brothers worked with their field-neighbours (for lack of a better translation of the expression "*hamara san-ji*") from the neighbouring village of Bapūli. They rented or "leased" the land adjoining the household, collectively working it and splitting the yield 50/50. A few Lohar and Rajput men from the financially stronger households held white-collar occupations, commuting to the larger cities nearby. For all larger landholders, choices made in the present were made to ensure that the future would enable the forthcoming generation to have the possibility of social mobility, especially via education. This was first and foremost provided to sons so that they could run the farm as a subsidiary source of income to (preferably) white-collar jobs. By 2013 there was one Lohar working as an accountant, one Rajput fire officer, one Rajput under training to become a police officer and one Rajput working as a cameraman for the local TV station, all males and all of them belonging to the three families with the largest landholdings.

These white-collar occupations paid better wages (especially those in the private sector, which were the most sought after), which led to trickle-down effects for their children, with respect to both building social networks outside the village and to pure social mobility, enabling parents to plan for private education. Education was regarded as a means to ensure a stable income in the future for most boys, and, thus, children of both genders below the age of 14 attended the public primary and middle school during my stay. A few children, especially girls from the poorer segments of the village, would complete their schooling at the age of 14, which was regarded as sufficient for a life dedicated to marriage and farming, but most children of both sexes continued schooling until the age of 16 (10th standard). Those with parents able to provide the financial support would continue through higher secondary until the age of 18. Some boys from these households also held college degrees in 2013, and in 2014, the first girl from the village signed up for a college degree. Government schooling is free, but by 2013 rural public schooling in the area was found inadequate. The government schools and colleges had a bad reputation, and most young parents in the wealthier households wished they could send their children to the private school in a town a few miles downhill in order to meet the requirements of the urban and contemporary society. (From my own experiences with Indian rural government schools, I have no reason to believe that this is an exaggeration.) The reality for marginal landholders, however, was very different.

Walking up the path to the houses at the highest elevation, the marginal and "landless" Rajput landowners' houses skirted the forest above. Many in these small houses at the top of the village self-reported to me as landless as they were not entitled to any arable land, but all of these people (of whom I was aware) had the benefit of "leasing" or "lending" small plots of arable land (often through lineage and inheritance). The plots under irrigation were perhaps small, but it meant that these families could utilise both growing seasons, which provided a minor relief to the household economy. However, they only provided scant fodder for livestock. The marginal landowners could thus not afford to stall-feed buffaloes and bullocks, and relied mainly on cows of the native breed and goats for milk and manure-production. Goats were practical in that they could graze in the forest in daytime, without supervision, but they gave little milk and were easy prey for the black panthers (*bagheera*), who occasionally roamed close to the village and killed a goat or a kid.

These houses were generally small, mostly with two rooms. In most marginal or landless households, the façade was of *pakkā* materials, but the backsides and flooring were still *kaččā* in clay/mud/dung with a wooden framework. Their courtyards were also small and often earthen. Their hearth (*čhūlhā*) would be located outdoors, often in a corner of the yard, with a temporary roofed construction to keep it dry during rains. Separate bathrooms were uncommon, and many would use the courtyard nook for their "bath" and the forest of the hills for "bowel movements". Even if most expressed a yearning for *pakkā* houses, especially because of the repetitive repairs that had to be done to the earthen constructions as the rains broke them down, the old houses with *kaććā* floors and back walls are, by far, the most temperate. Smaller and with fewer windows than their *pakkā* counterparts, they remain shady and relatively cool under the heat and during the cold weeks of winter. The *kaććā* buildings retain more heat, making the older houses a tad warmer than upgraded ones. With their cemented floors, larger windows and overall size, and higher ceilings, *pakkā* houses do not keep the warmth very well; they remain cold and damp in winter and unbearably hot during the summer. This was not a substantial enough argument to counter the convenience of a *pakkā* house, however, and striving to upgrade the houses to a *pakkā* state would constantly add to the household expenses. Another expense for some, was household water access.

As the access to the *kuhl* for drinking-water purposes was restricted based on both landholding status and elevation (the *kuhl* could only provide household tap-water for households lying lower than its base), the landholding families at the highest elevation would use *kuhl* water mainly for irrigation and laundry, but utilize the government well-water for drinking and cooking purposes. This water came at a cost, which, in 2017, was at approximately Rs. 2 per kilo-litre of water in a rural, domestic household (including a 25% bill fee) (Haryana Public Health Engineering Department 2017). Additionally, it was only available a few hours a day, which made the daily household chores more difficult than for their lower-lying neighbours in the village centre.

Poverty amongst the households of the main village, especially amongst the Rajputs, seemed to be mostly the result of the (continuing) decrease and split-up of landholding as per patrilineal laws of succession. Tension and quarrels in this segment tended to be about land, especially between brothers who felt unfairly treated when plots were partitioned. A father, according to custom, splits his land equally between his sons, who are left with smaller and smaller landholdings, unless one or more of them moves out, sells or changes occupations. This practice has left many households and families poverty-stricken over time, unable to gather the resources needed for social mobilisation through education and white-collar labour. The situation was far worse for another "landless" settlement, however, which is where caste comes into sharp relief. I wrote "landless" in quotation marks because the issue of landlessness is complex (I will return to this in Chapter 2). The marginal or practically landless landholders were of Rajput caste, entitled to utilise the *kuhl* through lineage and decent. The S.Cs of Rani Mājri, however, were not. For some (often a privileged and internationally savvy elite) caste identity is fading in some ways but persisting vigorously in others, such as in marriage. For others, caste identity is far from a matter of the past, and the unwillingness to share food, social life and space on the basis of caste identity still structures most aspects of social life in many villages like Rani Mājri. The relationship between the high and low castes could perhaps best be described as unequal, but not fixed. Sometimes, political and economic relations would also develop across or despite caste identities and were sometimes aligned or strengthened by them. As the configuration of gender, caste and class matters to such a degree for everyday practices in the village, a short outline will be provided here.

Caste in Rani Mājri

Caste is an endogamous category, which means that you are born into your caste group as well as into a linage, and in Rani Mājri there were three castes – Rajput, Lohar and Chāmar. In theory, at least, the caste system is pan-Indian (with regional variations) and that caste as *jāti* (lit. "kind", "species") is based on labour division (Lamb 2000:35; Fuller 2004:13,15), notions of ritual purity and impurity (Dumont 1980) and of auspiciousness and inauspiciousness (Raheja 1988a,b).

The Rajput caste of Rani Mājri were mostly farmers, and the major landholding caste of the village. By descent, the Rajputs were traditionally a warrior-caste, occupying a ritually superior position in the overall Hindu caste system. Their settlement followed the trajectory of the irrigation canal, the *kuhl*, as it made its way down from the hills, through the village and towards the fields. One woman from a Rajput household was working as a primary school cook and teaching assistant, but Rajput males, except those very few that could live off managing the largest farms, would commute to their jobs. Within the Rajput caste, the identity as farmer and landowner was highly underscored, and in this segment very few talked about ever moving out or giving up farming completely.

This was different amongst families of the Lohar caste, who for generations had benefited greatly from combining farming with trade, though their farmer identity was underplayed in relation to their artisan one. The Lohar caste of Rani Mājri are craftsmen, working with wood, and the only artisan caste of the village. By 2013, many Lohar families had complemented farming with carpentry one male worked as the shopkeeper in the local roadside shop, one worked as a lorry driver, and two women were caretakers in the village day care centre, for a small government salary. Some of these Lohar caste households had been quite successful and had bought large amounts of land. Their houses were roughly built along a well-trodden path skirting the fields to the south and the riverbed to the north, leading down towards the plains, but one Lohar family had purchased and built a house in the midst of Rajput houses. This had caused one of the few large quarrels in the village the previous decade, and, in 2013, the issue was still quite delicate. I found the focus on education and "office jobs" particularly strong amongst the Lohars, who, contrary to their Rajput neighbours, saw no reason to continue their fathers' farming practices unless they had to. This ability to leave farming behind to embark on more prosperous careers had left very few Lohar households in the same poverty as some Rajput families.

People from the Lohar and Rajput castes kept to their own separate communities in everyday social life and at more minor ritual events. But on other occasions they were, to a certain degree, mixed – both in housing and at larger village gatherings.

The third caste was the Chāmars or the S.Cs of Rani Mājri. The two other castes do not socialise with this one, except in government education. The caste name of "Chāmar" was seen as derogatory by both internal and external definition, and, thus, I have opted for the official state and internally used category of "Scheduled Caste" or "S.C." throughout my text. As small-scale Agriculturalists, they work with leather and dispose of dead animals, with a separate hamlet further south and further down the terrain. Neither of the two higher castes accepts any food or drink from S.Cs, and one is expected not to enter into relationships that involve physical touch, direct or indirect, between the caste groups.

In village celebrations or in other contexts where the higher castes eat alongside the S.Cs, the latter will be served and seated separately, and often after the two higher castes. The S.Cs keep a respectful distance and will only be seen in the village centre at times of traveling out of the area, as the only accessible road connection to the plains run through the centre of the village, or when sweeping the streets of the main village. One S.C. female did the washing at the schools; these kinds of occupations, considered defiling tasks for a Hindu, were compensated with small salaries from the government. Another young S.C. male took on additional work as tailor for the high castes, but mostly, such young men work as day-to-day manual labourers for cash or for part of crop yields when assisting larger farms in the district during harvests. Generally treated with polite contempt from both sides, there were of course exceptions. Some members of the S.Cs were better liked than others, either because of personality or skills they possessed. Similarly, the S.Cs would think more highly of certain high-caste people than others. Also, norms of high-caste etiquette, such as abstaining from meat, alcohol and cigarettes, were negotiable realities. The occasional indulgence in goat's meat or egg gathered most males from the central village in a jolly camaraderie from time to time (women from the same households, however, would disapprove). The same could be said for indulgence in alcoholic beverages or cigarettes, which were tolerated (but still not socially approved) for a man but not for a woman (except smoking, which elderly widows were known to do).

Of Hinduism in India, it is said that the manner in which Hindus playfully engage in ritual practices across conventional religious boundaries acts as a form of multi-faith "osmotic worlding" (Frøystad 2016). Thus, it could be argued to be so diversified in its expression that "any particular practice or belief may be contradicted elsewhere or denied by some Hindu" (Wadley

1977:113–114). Obviously, as in South Asian Hinduism itself, caste in practice is often more diversified than it is unified.

To explain why is a complicated exercise, and anthropologists have been ferociously debating whether the ritual opposition between purity and impurity is the structuring principle in the hierarchy of caste relations or not. Louis Dumont (1980) gave a thorough analysis of the Indian caste system, where the ritual polarity of purity–pollution was the central organising principle. His analysis, through debated and criticised over the years, has spurred many a nuanced contribution to the relationship between caste and rank, royalty and power. This especially affects the discussion around caste identity as founded in a structure of hierarchy revolving around *either* the ritual aspect of purity and pollution – and thus the pre-eminence of the Brahman (priestly caste) – *or* ritual aspects of power and distribution, thus giving pre-eminence to the Kshatriya (royal caste). Raheja uses the intricate norms of gift-giving (*dān*) in Hinduism to illustrate how caste and hierarchy play out in the daily lives of Hindus in rural North India. In Raheja's own fieldwork village of Pahansu, Uttar Pradesh, the dominant landholding caste gave *dān* (gifts) to all castes to ensure well-being, prosperity and auspiciousness for their families, harvests and the village as a whole – or to the village gods and deities to remove, or "make far", misfortune and evil (Raheja 1988a:511). In *dān*, argues Raheja, there's a "poison" (inauspiciousness), and in the act of giving *dān*, they are pivoting ritual capacity in a centre-periphery model. Raheja argues that it is not impurity that is given from a caste of high ritual status with *dān* but inauspiciousness. By giving away inauspiciousness, the landholding caste keeps his own family auspicious, partly explaining his dominance over the ritually purer Brahmins. This argument disagrees with the Dumontian centrality of purity and pollution to explain hierarchy and instead situates the transfer of auspiciousness and inauspiciousness (*śubh/aśubh*) as the central principle of caste hierarchy.

As the amount of benevolent and malevolent substances or potentialities are defined by one's caste identity, one thus avoids interaction with people from certain castes because they might, to varying degrees, contaminate one's own "body essence". This might become clearer when one approaches caste as notions of substance-codes through the work of Marriott (1990) or as a flow of "vital energy" (Raheja 1988a:506), but perhaps most illuminating is Zimmerman's interpretation of the Hindu body. Zimmermann (2014) argued that the Hindu body, at least through Sanskrit texts, consists of a network of channels or veins. Like a river, the network of channels starts out from where it left off in the karmic cycle, then changes its hue throughout a lifespan, aligning more or less with the three "body essences" and qualities (*gunas*): *sattva*, *rajas* and *tamas*. These substances have associated tempers, tastes and dispositions. In the priestly caste (Brahman) *sattva* – the purest, primary essence of the cosmos associated with knowledge and clarity – is thought to dominate. In the warrior castes (Kshatriyas, such as the Rajputs of Rani Mājri), *rajas* – the essence of action and heat – is thought to be dominant. In the Shudras (the servant castes), *tamas* – the most impure essence that can cause indolence, delusion and ignorance – is thought to dominate.

These essences are seen to fluctuate and relate to the body in various ways. A contamination of the body essence is observed when a substance of purer value comes into contact with a substance of poorer value, such as an S.C. directly or indirectly touching a Brahmin, as vividly illustrated in Lamb's (2000) fieldwork amongst Brahmins in West Bengal. Being the purest caste, a Brahman must be particularly careful about pollution from the substance-flow from lower, more impure castes, and after even an indirect touch (such as being seated feet apart but on the same mat with a lower caste individual), a Brahman would have to bath and cleanse himself or herself of the caste pollution (Lamb 2000:32). Peoples actions, inactions, who they are with and what they eat are seen to affect this balance of essences in their body. The preparation, sharing and timing of food are especially tied to notions of caste in India (see Daniel 1984; Sax 1990; Lamb 2000). For example, eating a fruit or vegetable with heating qualities when you are pregnant might cause trauma or illness as a "cool" body state is thought to be beneficial for the baby. There are also polluting substances, which include bodily emissions and waste matters (Raheja 1988a,b; Lamb 2000; Fuller 2004:15). The control of these emissions is thus made subject to quite strict social norms and regulations (food prepared by lower castes, female menstruation and sexual energy are examples of this).

Water, alongside fire, is considered a purifying substance in Hinduism, and the running water of rivers and channels is used for purifying one's body and preparing it ritually for performing worship. It is also used in rituals and worship directly. As we see in Sharma (1998) and Haberman (2006), in their respective analyses of the role of water in the rivers Ganga and Yamuna, the water of rivers is able to carry away human mortal sin and its associated substance, moving it away with its free flowing current. (Stagnant water, however, as Mathur [2001:25], and Rosin [2000] note, lacks such a purifying quality). Water, it should be noted, is extremely potent as a potentially contaminant vessel – an important point to note as access to *kuhl* water for irrigation and household activities was, in Rani Mājri, an issue of caste, though not exclusively so.

As with entitlement to water and land, caste status often, but not always, overlaps with class. This is, however, a rule with many exceptions as caste does not exist in isolation from other dynamics of power and identity. Occupations can be, and are being, pursued outside caste indications, which means that labour divisions do not signify caste belonging as much as they might have traditionally. Caste, however, still stands apart from other forms of stratification in the elaborate and ritualised practices that ordain and sanction the norms (Gupta 2005:410). Additionally, caste identity blends with the Indian state government practice of "affirmative action" in which quotas are set to increase the representation of marginalised caste and tribal groups: namely S.Cs, Scheduled Tribes and Other Backward Castes. Caste

thus effectively functions as leverage or an impediment in politics and finance, simultaneously blending and developing in conjunction with, or in
contrast to, local and global changes in the general norms of society. Still,
with few exceptions, the higher castes in the Shivalik Hills tend to own more
land and make more money, which enables more choices in life. Lower or
marginalised castes tend to own no or little land and thus more often get
caught in a spiral of unskilled labour and a hand-to-mouth existence.

Landownership, however, does not rely on income and caste identity
alone; it also depends on gender. Women in Rani Mājri did not own land,
and were often marginalized in the context of economic and political decisions in a household.

Women (and gender) in Rani Mājri

While it is important to note that gender as a socially constructed category that encompass more than the female perspective, my experiences as
a woman amongst women informed my understanding of their marginali-
zation in the public sphere. With regards to gender, the logic of the caste
system explains some of these gendered practices. As mentioned above, one
important principle underlying the caste system is the contrast between
ritual purity and pollution, and women are, at least at certain periods post
puberty, ritually impure. For example, fertile women in India cannot attain
complete ritual purity as menstruation and childbirth are bodily practices
that are considered to be severely polluting. After the onset of menstruation
especially, girls and women experience cyclical regulations on their mobility
as menstrual blood is considered a severely polluting transfer of substance.
General travel out of the village would thus be discouraged, and particular
areas in the village had to be avoided. Restrictions on entering the kitchen
in Rani Mājri households were, however, not as rigid as those I had experienced in Rajasthan – nor was cooking – but touching of water vessels had
to be avoided. Crossing the field at certain stages of cultivation or generally
entering or trespassing the sites of deities when menstruating could have
devastating results as it would upset the gods and destroy the crops. Other
deities' shrines, like Hanuman and Panch Pīr (see Chapter 4), can never
be visited by women. Post-puberty girls also never sit on windowsills, in
doorways or in other zones of transgression. The female body is also seen as
more fragile and susceptible malevolent substances than males, as females
according to Lamb (2000) are considered more "open" than men, and thus
vulnerable to polluting essences, especially heat and fluidity. Additionally,
as Fuller (2004:22) points out, Hindu women are not thought to be "twice
born" (i.e., they do not undergo an initiation rite in early adulthood) and as
such are ritually inferior to males. Due to these ritual conceptions of essential vulnerability as well as potential destructivity of women, their everyday
practices were, to a larger degree than men, controlled and her role as independent individual strongly subdued.

In North India, the role of maintaining family honour is a particularly salient aspect of being female. Irrespective of age, her "misbehaving" would reflect badly not only upon the woman herself but on her husband, children and extended family. Another difference, related to the former, was how women were expected to alter their appearance, with dress, acts, speech and movement, according to where and with whom they socialised. This would be influenced by age or status (as daughter, sister, mother, wife, etc.) but also by bodily cycles: for example, the practice of veiling. The traditional *salwar-qamīz* suits – long pants and tunics – are accompanied by a shawl, called *chunni*. When young and unmarried, the chunni is draped in front of the chest with the tails flowing down the woman's back, never to be used for veiling in her home village. Married women will always use the chunni to veil their eyes and face in the company of men, but as years pass and she reaches maturity, a woman in the company of her closest family will seldom fully veil, instead wearing her chunni casually covering her hair.

Females are in general strongly discouraged from engaging with males to whom they are unrelated by birth. They seldom travel, except for extraordinary situations (such as births and deaths) or for ritual occasions or festivals, and never without parents or a chaperone. Necessary travels over shorter distances related to work or studies would be accepted if she was traveling in a group of trusted friends. When married, a wife should not address her husband by name, and it is preferred that she not speak to him unless spoken to. She would also never speak to a man in her in-laws' house or to an older women unless asked to (and never with a raised voice). She would be served last and keep to the periphery or even withdraw herself completely to the kitchen or her bedroom if strangers or guests of any standing arrived. With time, she will eventually be allowed to partake in outdoors work, such as farming and tending to the cattle. If her children are young, a grandmother or any other woman confined to the household for any reason (such as being pregnant, post-labour, or in any other way unable to work) will tend to her children. If a family does not have someone in the house, the village *anganwadi* (day care) might be used whilst farm and husbandry duties are tended to.

All married women in the village were allowed to move about according to the rhythm of work, weather and landscape. Adult women could normally walk in the forest alone and travel wide distances for work in the field but would seldom make social visits to other women. If a woman is the oldest – or only – daughter-in-law (*bahū*) in a joint household, she might, as she ages, take on the position of "housewife". This English expression is used to honour the woman with the practical family responsibility, allowing her to oversee what needs to be done within the house and be responsible for the preparation (or delegating the preparation) of meals. Women have little time for networking, public affairs or friends. A good "housewife" here is characterised by how well she tends to her household, which means working in the house, the fields, or the forest, no matter the time of day, the day of the week or the season. A good "housewife" would, with age and leverage, be able to delegate the more tedious labour to her own daughters-in-law and meet her old age with more ease. But that is a position hard-earned. To become a "housewife" in the village of Rani Mājri is to have little time for anything but work. As well as doing what any other woman would do – her own, her husband and children's laundry by hand; sweeping the floors; feeding and tending to buffaloes, goats, bulls and/or cows; fetching fodder; sowing and harvesting; and otherwise tending to her children, husband and parents-in-law – the "housewife" is always the first to rise in the morning and quite often the last to go to bed at night.

The concept of the Hindu body does illustrate some of the structures that shape the gendered body, however changing and contested it may be. As Lamb (2000) and Sharma (2008) rightly point out, a woman's status changes several times during her lifetime, as do the relations between men and women, and those between women themselves. Gender, then, is a contested and rich category that coexists with all other hierarchical structures, such as caste and class. Since Berreman observed it in the 1960s (Berreman 1978), male work-migration from the hilly regions to the plains has been an increasing trend. By 2013, the trick for the larger landowning families of the village in contemporary North India was to find the right balance between farming for self-sufficiency and farming for income. This balance could be achieved by producing several sons, so that one or more could benefit from education and paid labour. This allowed for at least one son to keep the farm going. If the farm was sizeable, it required the sons to marry wives who could dedicate themselves to farming. If the farm was marginal and produced too little to make a profit on sale, the husband would pursue wage-labour, leaving the wife alone with the farm, the livestock and/or children and elderly in-laws. In the case of Bhagwati and Bhupati's household for example, the three daughters-in-law were all proficient Agriculturalists, and although they would very much like to educate their own daughters, they required a hard-working daughter-in-law (*mehnatī bahū*).

This has left women of landholding families in a difficult situation, and the landless S.Cs are often worse off. Women are more often deprioritised in education, and their participation in issues involving their own, and/or their children's well-being, is close to non-existent. By 2012–2016, the lack of literacy amongst women aged above 40, and the lack of higher education in general, made them aware of their "unawareness". Younger women, particularly, were largely cut off from the flow of information about politics and the economy. Decisions around large investments, loans, etc. would seldom (or never) be discussed with wives but left within the male group. Changes occurring in household labour division have not really benefited the females but, rather, have left them with more hard work to do.

References

Agarwal, A., and Narain, S.
2000   Redressing Ecological Poverty Through Participatory Democracy: Case Studies from India. *PERI Working Paper Series* (36): 29.

Arya, S.L., Agnihotri, Y. and Samra, J.S.
1994   Watershed-Management: Changes in Animal Population Structure, Income, and Cattle Migration, Shiwaliks, India. *Ambio* 23(7): 446–450.

Berreman, G.D.
1978   Ecology, Demography and Domestic Strategies in the Western Himalayas. *Journal of Anthropological Research* 34(3): 326–368.

Daniel, E. V.
1984   Fluid Signs: Being a Person the Tamil Way. Berkeley, Los Angeles, London: University of California Press.

Dumont, L.
1980   Homo Hierarchicus: The Caste System and Its Implications. University of Chicago Press.

Fuller, C.J.
2004   The Camphor Flame: Popular Hinduism and Society in India. 2nd edition. Princeton, NJ: Princeton University Press.

Gupta, D.
2005   Caste and Politics: Identity over System. *Annual Review of Anthropology* 34: 409–427.

Haberman, D.
2006   River of Love in an Age of Pollution: The Yamuna River of Northern India. Berkeley: University of California Press.

Haryana Public Health Engineering Department
2017   Rural Billing. *Government of Haryana*. https://biswas.phedharyana.gov.in/WriteReadData/Notice/CurrentTariff.pdf, accessed July 5, 2020.

I.I.S.W.C.
2012   Pers. comm. Group interview. Anonym. Central Soil & Water Conservation Research and Training Institute, Research Centre, Chandigarh.

Joshi, B.R.
2017   Sustainable Livestock Production in the Mountain Agro-Ecosystem of Nepal. Food and Agriculture Organization of the United Nations. *Agriculture and Consumer Protection Department*. http://www.fao.org/docrep/004/T0706E/T0706E04.htm, accessed February 2, 2017.

Lamb, S.
2000   White Saris and Sweet Mangoes: Aging, Gender and Body in North India. Berkeley and Los Angeles: University of California Press.

Marriott, M.
1990   India through Hindu Categories. New Delhi, Newbury Park, London: Sage Publications.

Mathur, N.
2001   Myth, Image and Ecology. *Indian Anthropologist* 31(1): 19–28.

Ministry of Agriculture & Farmers Welfare
2019   Categorization of Farmers. *Government of India*. https://pib.gov.in/newsite/PrintRelease.aspx?relid=188051, accessed March 10, 2020.

Narain, S.
2014   Opinion. How the Monsoon Has Changed. *Business Standard India*, September 7. http://www.business-standard.com/article/opinion/sunita-narain-how-the-monsoon-has-changed-114090700762_1.html, accessed March 20, 2017.
Raheja, G.G.
1988a   India: Caste, Kingship, and Dominance Reconsidered. *Annual Review of Anthropology* 17(1): 497–522.
1988b   The Poison in the Gift: Ritual, Prestation, and the Dominant Caste in a North Indian Village. Chicago: University of Chicago Press.
Rosin, R.T.
2000   Wind, Traffic and Dust: The Recycling of Wastes. *Contributions to Indian Sociology* 34(3): 361–408.
Sax, William S.
1990   Village Daughter, Village Goddess: Residence, Gender, and Politics in a Himalayan Pilgrimage. *American Ethnologist* 17(3): 491–512.
Sharma, A.
2008   Logics of Empowerment: Development, Gender, and Governance in Neoliberal India. Minneapolis; London: University of Minnesota Press. Retrieved September 1, 2020, from http://www.jstor.org/stable/10.5749/j.ctttv9xh
Sharma, R.D.
1998   Sacred Immanence: Reflections of Ecofeminism in Hindu Tantra. In *Purifying the Earthly Body of God: Religion and Ecology in Hindu India*. Lance E. Nelson, ed. Pp. 89–131. SUNY Series in Religious Studies. Albany, NY: State University of New York Press.
Shira, D.
2015   Monsoon Season: India's Real Finance Minister? Article in *India Briefing News*. http://www.india-briefing.com/news/monsoon-season-indias-real-finance-minister-10714.html/, accessed March 20, 2017.
The Tribune
2013   The Tribune, Chandigarh, India - Ludhiana Stories. *http://www.tribuneindia.com/2013/20130614/ldh1.html*, accessed March 20, 2017.
Yadav, R.P., Singh, P., Arya, S.L., Bhatt, V.K. and Sharma, P.
Wadley, S.S.
1977   Women and the Hindu Tradition. *Signs* 3(1): 113–125.
2008   Detailed Project Report: For Implementation under NWDPRA Scheme, *Central Soil & Water Conservation Research and Training Institute, Research Centre, Chandigarh*. Ministry of Agriculture, Govt. of India, New Delhi.
Zimmermann, F.O.
2014   The Jungle and the Aroma of Meats: An Ecological Theme in Hindu Medicine. In *The Anthropology of Climate Change: An Historical Reader*. 1st edition. Pp. 67–80. West Sussex, UK: Wiley-Blackwell.

2 Waterworn

Śharad

Śharad ritu, the *"early autumn"* season in the village of Rani Mājri, is a season of growth that lasts through the lunar months Bhādoṅ (mid-August to
mid-September) through Āsin (mid-September to mid-October). These early
autumn days are warm, but the clouds in the sky do not always carry rain. As
the monsoon crops have been in the ground for a while, the fields are now a
wonderful mosaic of different shades of green. In the now lush forests above
the village, the guava ripens on the intermittent trees, and the water in the kuhl
runs forcefully from the hills.

One early autumn morning, I see Avani, a neighbour of ours, in her courtyard, about to fill a basket of manure. She is sweeping the little there is from
her doe, couple of kids, and her skinny old Indian-breed cow. I like spending
time with Avani, but she seldom finds the time to sit for a cup of tea. Thin,
intelligent and agile, Avani is a Rajput woman in her mid-forties who works
all the time: either tending to her land and animals or at her part-time job as a
caretaker at the local primary school. Her husband is not well and often unable
to work because of an "illness" in his head (my poor understanding of the local
dialect sometimes made precise diagnostics difficult to convey). The medicines
he need squeezes their household budget tighter every year. Her salary also
covers the education of her two sons, both away at government colleges. Await-
ing their return, two small, cemented bedrooms were constructed recently, next
to the old and small two-room house, made of cow-dung and clay. Soon, they
would come home, marry and take on jobs. Avani almost cannot wait. As I
have an hour just for myself for once, I ask if can join her to go to the fields with
the manure and perhaps help her weed for grass fodder.

Arriving at their irrigated patch, the one with tomatoes ripening, she empties
the manure onto a pile to be composted and later added to the irrigated field.
She then hunches down to weed; the grass is good fodder for her pregnant
cow. I watch her work with that blend of care and efficiency that makes you
recognise skill, and then hunch down to do the same. I am as a child beside her.
Thinking that all weed is simply weed, I almost cause a disaster by not recog-
nising one poisonous plant from another. "This one makes the milk taste sour",
Avani explains while she throws it away, "and this is good for pregnant cattle!"
I feel utterly useless, especially after I grab a handful of what in all respects

must be called ordinary grass, on the wrong side of an invisible line between to plots. Avani instantly corrects me: "That is someone else's grass, leave it!" Having caused enough trouble, I instead ask Avani to show me her fields, if she has the time. She does, surprisingly, and we both embrace this sudden opportunity to roam about. Enthusiastically, we jump, balance and skip southward and downward along the kuhl. On the way, she comments upon the various plots. She tells me who they belong to and points out how well the land has been tended, whether the pebbles from the previous monsoon have been removed and whether the manure has been collected neatly. "Look, that big field over there, with the store-bought fences, look, that belongs to the Lambardar", she says, raising her eyebrows in that characteristic manner. "Lambardar" refers to a hereditary status of a large landowner who, in the past, had certain policing rights in the village. In Rani Mājri, the Lambardars were large, landholding Rajputs, and the Lambardar family was the only one who could afford fences. A bit further downhill, we pass a small, stilted house where those who cannot afford fencing must sleep (or wake) to protect their crops from wild deer and boars emerging from the forest in the hills to indulge.

The further downhill we go, the more the stones that encircle the terraced fields vanish. Here, the land is rain-fed, not irrigated, and the plots are no

Figure 2.1 A shelter on the fields where the farmers can protect the crops from bovids due to a lack of proper fencing.
Source: Author.

longer terraced but follow the topography of the sloping hill. This makes it difficult for me to distinguish one field from another; however, Avani does. "Look, that's a Harijian[1] field. See, how they have not even bothered to weed and pick the stones away? They are so lazy!" she exclaims, referring to a group of S.C. people living outside the village centre. I stop to look at a structure of stone piled up on top of each other, sparsely decorated with some white paint. It is for a deity belonging to the Harijan, Avani explains, but she cannot remember the name of it. She's eager to get on – "Just wait until you see our fields!" she exclaims. Now, having almost reached Rani Mājri's neighbouring village of Khot, we traverse through some bushes ("a bamboo, look! Isn't it marvellous?! It has grown sooo much since last year!"), and, finally, we are standing on Avani's household's rain-fed land. No pebbles cover the ground here; it does indeed look well-tended. The plot is small and steep, however, as the land rises up towards another ridge to the south. They grow lentils and horse gram here, she tells me, which require less daily supervision. "But not this year", she says, "this year we have sown another land. Next year, here". A glimpse of a peacock in flight just a few yards away prompts sudden silence until it is beyond the horizon. "A peacock, on my land…" she whispers dreamily. "It is a good sign". She nods, and we hurry home.

During these early autumn days, recruiters from the private schools and colleges on the plains are welcomed into homes and courtyards. In couples, they approach mothers and fathers with colourful brochures and pamphlets. As schools are to reopen after the holidays, dreams and hopes for the future are revived on behalf of the next generation. Most often ambitions towards higher education involve boys only as girls are destined to be farmers' wives and thus not in need of higher education, but there is a growing awareness that times are changing. Parents know that in the cities, girls study, work and earn money. But the risk of "over-educating" your daughter with ideas far above her station might quickly leave your cherished daughter in misery and woe if her in-laws require her to keep to their houses and farms. Some of these dreams and hopes for the future are taken to the deities as many make religious journeys during this relatively calm season of growth. Along the main roads, tens of thousands of men dressed in orange, the devotees of Lord Shiva, tread along on a pilgrimage to Haridwar to carry water home from the holy river Ganga.

The growing season is a particularly auspicious time for tending to relationships with the lineage or individually chosen deities, and relationships between people. As farm-work is now manageable, children who come home from school have time for rest, homework, needlework and television soap operas. Married women of all ages will be able to take time off from work to visit their family in their natal homes (māykā kā ghar) now, especially for the festival of Rakṣā Bandhan (also called Rākhī, lit.: Bond of Protection). As one moves from house to house, girls and women bring red wristbands, a sacred red thread, to tie on their brothers' wrists, and, in return, he offers her his patronage and protection. Red bands are also playfully tied around the wrists of cherished cousins or friends. The atmosphere of the festival is relaxed and

joyful; it is a day of fun and playfulness, where one enjoys time with family and
special foods that are prepared for guests, like the pakoṛa – *a salty and fatty*
snack made of fried onions. The pleasant temperature allows boys to play in
the street after school: a simplified version of basketball or cricket, with an old
bat and a cricket ball made of old socks.

At the onset of the month of Āsin, many women fast and visit the Durgā
temple in Dhamra. This temple is a very popular temple for several Hindu
castes, as well as Sikhs, five to six kilometres down the hill from Rani Mājri.
Women in beautiful garments would, during these days, be seen walking in
small groups from the villages and hamlets in the vicinity to "see the mother".
To worship Durgā at the Dhamra temple during the nine-night Navarātri fes-
tival of the season,[2] was regarded particularly auspicious. In Dhamra, the
Goddess could be addressed "in all her forms and manifestations" (Rodrigues
2003:16) and the temple thus attracted people from the entire region, both rich
and poor, high and low caste.[3] In the entire district, "Jay mātā dī" (victory to
the mother) could be read on just about every auto-rickshaw, and, amongst
the higher castes, Durgā as Sherawali mata *("Śerāṁvālī Mātā"), or "she who*
rides a lion" was particularly revered. She was a mighty ally too. Durgā is, in
Hinduism, the embodiment of shakti – the energy, force or the "dynamic power
of the universe" (Kinsley 1986; Erndl 1998:176). Uncontrolled, she is mighty
and potentially dangerous, but controlled in one of her forms (such as Parvati,
the devoted wife of Shiva), she can provide a benevolent gaze for the household,
in which she is approached for personal concerns to cure diseases, help people
in distress, and so on. A few weeks after Navarātri, the villagers also perform
the "ancestor spirit" pūja, where girls are introduced to their "atma shanti".[4]
I never witnessed this, but I was told that during the pūja, the parents would
wash the feet of all the girls of the household and give the "atma shanti" the
names of their girls so that it would give them protection. A pandit[5] would visit
each of the households of the village to partake in the proceedings, holding one
pūja for each household every day for a period of approximately 16 days.

As we move into October, it is late autumn in Rani Mājri. The maturing
crop adds shades of yellow and orange to the mosaic of green. A layer of dust
now covers the leaves of the forest, dulling them a little. The kuhl carries less
water, but it trickles steadily through the village and down to the fields. The
season's latest ritual is the Karvā Cauth, *when married women follow a strict*
fast (nothing but air should pass the lips and not even water is drunk) between
sunrise and moonrise for the prosperity and longevity of their husbands. This
is a popular ritual not only among the village women but also among the urban
female elite, who apply beautiful, hand-drawn mehndi *(mehaṁdī) on their*
hands. The paste made of dried henna plant dries to reveal the temporary tat-
too; harvesting time is around the corner, and work again picks up speed.

....

For Avani, whose path we followed above, everyday life and her hopes,
dreams, fears and ambitions are all entwined in the future of the scanty forest, dusty streets, irrigated land and large areas of sandy hillsides that have

collapsed into ravines of sand and stone. Alongside destructive wildlife, water was a source of constant worry for all villagers. While too-intense monsoon rains equal landslides, damaged crops, siltation and erosion, drought equals less nutritious and diverse foods, less milk, more debt and more hardship. Rani Mājri's landscape is in a way waterworn – shaped and reshaped again by water. The rain pours from the sky and runs down the hills, into the riverbed, the soil of the forest and the fields. Through seasons, years, water with wind, frost and sun, the landscape may look stagnant, but it moves. Its exterior – the forests, meadows and gardens – fluctuate with growth and decay, in a paced rhythm, one of familiarity. Earth and stones move slowly, however, altering the landscape in rhythms so sedate, they are unfamiliar to us. The availability of water has played, and continues to play, a central role in structuring the past and present of life in these hills. To walk along the *kuhl*, as I did with Avani, is also to walk through time, place and social relations. Through its periods of scarcity and abundance, the *kuhl* has carried water to irrigate these small terraced fields, in addition to providing water for drinking, household and ritual activities for hundreds of years.

With the cutting of trees, the maintenance and development of the *kuhl*, the expansion and intensification of farm land, the development of roads and networks, the grid of cables and dams and mines, the fencing of protected areas and the growth and expansion of markets, industry and townships, the landscape has also been shaped and reshaped by humans, animals and buildings. In conjunction with national socioeconomic development planning, several of the hill villages have become connected – although peripherally – to political campaigning, education services, health services, electricity and water-grids, telephone lines and media broadcasting services. This development has, especially after the 1970s, been closely related to international, national and regional initiatives to protect the hills from environmental deterioration. Controlling and managing water has thus carried the promise of a decent yield, or the omen of failure and hardship, from the time of the first settlements in these hills until today.

Being so central to ritual, social and economic life, the prime concern of the villagers that I noted during my stay was, surprisingly, not related to water. Water was unpredictable, due to the variations of the monsoon and the winter rains. The constant tension between having too little or too much water, as well as the variable accessibility of pure drinking water, made the thought of water a constant presence in the village. Water and its management are thus closely related to managing, allocating and utilising all that which it enriches, the soil and its bounty – whether it be forest, weed or crops. How these resources are shared in terms of caste, lineage, class and gender relations makes structuring this chapter around the distribution of water quite revealing when it comes to social relations. This chapter will show that the village is far from a uniform unit and that the differences that have persisted, or even increased, between certain groups have historical, cultural and political roots. These differences make for an uneven playing field when large international actors, regional government schemes or projects, private NGOs or scientists attempt to develop, conserve or even enlighten the villagers on the changes they, willingly or not, are taking part in.

Becoming Rani Mājri: a *kuhl* story

The *kuhl* began its pathway into Rani Mājri territory in 2013 as a small river or stream at the junction of two ridges. It is derived from a perennial river that runs down from Himachal Pradesh territory and down into a large, but mostly dried-out riverbed. The size of the riverbed indicates that the river was larger before, but no one in the village could remember it flowing to reach its banks. In the monsoon season, there is now barely enough water here for a small waterfall, a popular place for the youth to pose for Facebook profile pictures.

A lot of the water is harvested before it reaches the waterfall, however, as the stream pours from the northern hills into a small, sub-surface tank. Here it is diverted into two systems. One runs along the river-bottom in a pipe to provide *kuhl* water for the villages below Rani Mājri, pouring into their traditional *kuhl* systems. The other runs towards the Rani Mājri *kuhl* system as an open water channel (under the time of fieldwork this was a construction in process), with an adjacent, smaller pipeline for drinking water running along its side. In 2013, the Rani Mājri *kuhl* was still open – and accompanied by a narrow, stony path – the water channel cut along a steep valley side. Smaller footpaths diverged from the main path here, one leading upward to a tiny village of Brahmins, which I never had the chance to visit, another downward, to the Rajput cremation grounds (*shmashāna*), which I was never allowed to visit.

On its way southward towards the village, the water pours through two recently constructed cemented and open water-harvesting tanks, where it can be halted and stored for dry-season use (or by young boys for a cooling bath in warm weather). Eventually, it passes through a small mango grove, where a ruin of a brick construction marks the entrance to the village from the northern hills. Apparently, the ruin used to host the village deity Khera Baba, but after a large landslide occurred just next to it (I did not record the year this happened, but I was left with the impression it was sometime after the 1990s), a new temple was constructed for the deity on the other, southern side of the village. The path here becomes solid (*pakkā*) in this case, i.e., cemented. This forces the *kuhl* underground for a bit as it continues towards the fields, sometimes in the open, sometimes diving under a *porcha* (the small courtyards that front every house) as it passes through the village itself. At the lower end of the village, the *kuhl* resurfaces in an open space at the centre of the lower village in a *cul-de-sac*. Across the open space, just next to the Hanuman temple, the water plunges down into a system of numerous small channels in an intricate network, manually opened and closed to irrigate patched fields below.

The *kuhl* has not always looked like this, neither have the fields. When it was first dug, the *kuhl* was an open earthen channel, carrying water only when in abundance. For centuries, the water channel was maintained, repaired and improved by low-caste daily labourers, supervised by higher-caste landowners, blessed by the deities and financed through gifts; donations; kings; and, post-colonial rule, the Indian government. To walk along the *kuhl* system of Rani Mājri, then, is akin to walking along the village's history.

One day, the retired village Lambardar, an elderly Rajput man with no teeth and a wide grin, invited me and Prakash's daughter for a cup of tea. Living in a large house with his much younger wife, two broad-shouldered and sharply cut sons, and their respective wives and children, his house was one of many rooms, one of which hosted the village's only water heater. When I asked him about the history of the village, he, in a broad Pahari dialect, told me enthusiastically about the days of his youth. I am sure much detail has been lost, and more would have been lost if not for the patience of Prakash's daughter in attempting to recount his words to me in a mix of Hindi and English. But, with added detail from I.I.S.W.C.'s final project report for the watershed project (Yadav et al. 2008), other elders in the village (born in the 1930s and 1940s), as well as history books and colonial gazetteers, this is what I gathered of Rani Mājri's place in history.

Time beyond living memory

Around the year 1600, when the Baroque cultural movement was dominating the European peninsula, the Americas were slowly being colonised, the Ming Dynasty in China was collapsing and the Sikhs were beginning to rise to power in Punjab, the first earthen water channel, or *kaććā kuhl*, was dug by settlers from the hills in the north. During the 1600s, this area in the lower Shivalik Hills was most likely included in the Hindu kingdoms of Nahan and, later, Sirmaur, now a district in the state of Himachal Pradesh. Located in the lower-lying regions of the Shivalik Hills, the village in the 17th century bordered scattered townships of the plains below, just as it does in 2013, but the townships were far from as populous and were ruled by the Muslim Mughal Empire.

A family from the Rajput caste migrated to this area from the mountains above Rani Mājri. In the village from which they originated, there had been many sons and no land left to farm. Probably settling down around the same period were the crafting caste of Lohars and the S.Cs, although I have no more specific record of when they settled here other than it being "a very long time ago".

The first period of settlement, access to water for irrigation and drinking/household activities would have remained solely dependent on the amount of rainfall, and so the citizens dug a channel to lead the water from the river to the north onto fertile land. With irrigation, the water was allocated between the settlers, who built the first houses high in the terrain, above the fields, with the backs of the houses skirting the forest. Even today, the central village of Rani Mājri appears to cling to a hillside so as to avoid using up any land usable for cultivation. They had small houses with wooden frames and floors, walls and roofs made of clay and cow dung. Exposed to the sun and the rain and the heat and the cold, the houses were constantly rebuilt, and the village itself "shifted" further down the terrain at some point.

According to the Lambardar, also "many hundred years ago", the entire village of Rani Mājri and its land was given away as dowry in the marriage of a hill-king's daughter. Apparently, the princess had in fact lived in Rani Mājri with her husband for some time. The village got its name from this princess, and the toponym "Rani Mājri", translated as "the abode of queens", is partly derived from this origin story. This seems to have been a relatively ordinary practice in the region, as Mark Baker (2005) found that that 14 out of 19 pre-colonial, state-sponsored *kuhl*s in his study from Kangra Valley in the bordering state of Himachal Pradesh, were named after the king (*raja*) or queen (*rani*) who built them. Then, "sometime later", there was a plague that made the entire royal family flee to Nahan. Just before the death of the Nahan king, the Lambardar told me, he gave the land – of which Rani Mājri was part – as mortgage to the king of Patiala.

This transaction of the land could have happened sometime after 1691, which saw the first king of Patiala rise to power – the Jat Sikh Baba Ala Singh. The Jat Sikhs were, to my knowledge, Hindus that converted to Sikhism in 17th- and 18th-century India (Rajadhyaksha 2017). As a minor kingdom, the kings of Patiala would constantly negotiate alliances between Sikh-dominated areas to the west, the southern Maratha empire to the south and repeated invasions from the Mughal Empire. In the late 18th century, when the British East India Company initiated what became British colonial rule in India, the gazetteers – geographical dictionaries of a sort – documented British advances into the northern frontiers.

By 1805, the British had travelled northwards from their early establishments along the coast and "ceded and conquered" a relatively narrow passage, following the river Yamuna north from Delhi (Cunningham 1883:24). Four years later, in 1809, they reached the Patiala kingdom. The area was a strategic location as an entry point to the Himalayan hills and the (soon to be) colonial summer capital of Shimla (Gazetteer of the Simla District 1888–1889:27), and the British soon engaged in a war. This was not with the king of Patiala, who continuously sided with the British throughout colonisation but, rather, with the Gurkhas. The Gurkhas were Kshetri-Brahmin Nepalese elite from the northern hill states (Pemble 2009:374). After the Gurkhas had been defeated in the two-year "Anglo-Nepali war", certain parts of the northern provinces were now brought under British rule. The king of Patiala, however, kept some areas under his princely state, with internal autonomy, and among this was Rani Mājri.

According to the *Ambala Imperial Gazetteer* of 1909 (pp. 276–277), the chief crops grown in these hills were horse gram, a small legume typical in dry land agriculture, but also wheat, ginger and *kachālu* (a colocasia version of the taro plant) – crops that thrive in irrigated fields. The availability of maize and fresh greens and vegetables, as well as surplus produce for sale, seems to have been marginal in the area, which was, by 1909, sparsely populated. A nearby hill station reported suffering badly from cholera, fever and smallpox, and the village of Rani Mājri was probably no different. There are few descriptions about social life or the land itself in the hills around Rani Mājri in the early colonial gazetteers, maybe because the British themselves seemed more occupied with inculcating the young, new king of Patiala, Bhupinder Singh, known for his love for cricket, women, alcohol and extravagance (Grewal 2004:654). As the colonial empire grew, the peripheral villages too, fell under the governance of the British-Indian economy.

Time remembered

Throughout the era vaguely remembered by the eldest in the village (post-1930s) the village appears to have been in an impoverished state. According to the watershed project report from Yadav et al. (2008), only six families lived in central Rani Mājri by 1930, and, at that time, villagers paid rent (*lagan*) to the Patiala king: 1 rupee for irrigated land and 25 paisa for rain-fed land, collected by the village Patwari (village accountant), the Lambardar of Rani Mājri. The elders of the village recalled the mid-forties as a time of radical change, marked by poverty, hardship and strife. One elderly farmer said that, back then, people had "almost nothing, barely clothes". Bhagwati, an elderly Rajput woman who married into the village at the age of 12, also recalled the 1950s and 1960s as a time when people only worked, and life was hard.

> Back then, all the houses were *kaččā* [here: constructed from clay/dung/mud and wooden frameworks]. No proper road connection went to the village, there was no bus, no autos [auto-rickshaws], no cars - it was just a path and you had to walk if you were going somewhere.
>
> – Bhagwati

Disconnected from the grids that have now developed in the more urban areas of India, with electricity, trade and communication, the village appeared to be on the political and economic periphery. As the village is geographically located on the hilly and rural side of the river Ghaggar, on the opposite side of townships and cities like Pinjore, reaching a town large enough to host a well-sized market or a doctor would, in that era, take a day of travel by foot. To get drinking water, women from the village would walk the narrow path along to the northern riverbed, where small perennial springs surfaced from the sandy ground, in order to refill their water vessels in the drought of summer.

The houses of the village were small and still made of wood, clay and dung. They had a small nook for goats and an outdoor hearth. This is how the elderly remember the houses to "always" have been. For hundreds of years, village houses would expand when there were times of plenty and collapse in times of heavy rain and scarcity. That was what houses had "always" done. Soon, two forms of new technology would cause a definitive break with the "old days" – the introduction of electricity and the Green Revolution together radically transformed everyday life in the village.

During the next few decades, much would improve for the villagers in Rani Mājri, but they would also face several new challenges and problems. Once British rule ended in 1947, the kingdom was officially subsumed under the Indian Republic. The new state continued with a strong control over forests and land, and, to achieve economic growth, large areas of forested lands were converted for other uses (Baviskar 2001:249). Coupled with an international post-Second World War urging the development of the not-yet industrialised countries of the world, the concept of development would affect this state's rural policies. Although those processes normally associated with development (education, technology, economic growth, etc.) were slow here, compared to those in other areas in the plain's region, development eventually materialised in Rani Mājri, often aided by the villagers themselves.

In 1956, for example, the first initiative for rural education was initiated by a villager who started a small school in his private house. This initiative was apparently so popular that villages in the vicinity collected money to give as a salary to the self-appointed teacher. Girls were not allowed to attend tuition at the time, but the boys who, in 2013, made up the politically active and decision-making segment of the population received their first years of schooling here. In 1960, the first radio was also bought by a villager – it was a battery radio as no electricity was available at the time – and in 1961, the government established a primary school in Rani Mājri, and so girls too were allowed an elementary education. The school, a small, cement building with a tiny courtyard in the middle of the village, included the first to sixth grade. Males born in the 1960s thus usually had six to eight years of elementary schooling, while the women of the same age I recorded ranged from having no schooling at all to having four or (in rare cases) six years.

The village was finally connected to the electricity grid in 1976, which is rather late in an Indian context. For example, electricity had already arrived to India via Calcutta (Kolkata) in the 1890s (Dash 2009). ("Late" is however relative; by 2009, there were still villages in the Shivalik Hills not attached to the grid, according to the International Business Publications 2011.) Bhagwati remembered well the days without electricity and told me that those were indeed darker times: "Before we had to use cotton wicks burning in oil for light, but it gave off so very little. But when electricity came, everything got so much better – now we had light!"

Only one year later, in 1977, the village was connected by a partly paved road to larger cities in the vicinity and, subsequently, to the Delhi-Shimla Highway. This made transport and communication to and from the village easier. Over a few years, the houses of those who could afford it now slowly became solidified (*pakkā*) with the addition of bricks, cement and steel. Added on here, replaced there, the houses changed and the husbandry changed too. Goats were slowly replaced by cows – animals that would give more milk but demand more space as they had to be stall-fed, now made possible by advances in building styles. Old bedrooms were now, by those few who could afford to, converted to cattle sheds. In the house in which I lived, faded and torn posters picturing dancing deities on the earthen walls served as a faded memory of its being the bedroom of a young, newlywed couple in the mid-1980s.

Simultaneously, the hill region's precarious ecological state was discovered, which encouraged the development of the rural Shivalik Hills. During the 1970s, India's government echoed international concerns over the environment and adopted a conservationist orientation (Baviskar 2001:249). This occurred as the urban town planners of Chandigarh acutely realised that the prosperity of the plains region was in many ways dependent on the social and environmental well-being of the hills, as it was discovered that the drinking water reservoir of Chandigarh, Sukhna Lake (constructed in 1958), had for years been filling up with silt from the denuded forests and hills above the city. The Chandigarh government had, by the 1970s, already used US$200,000 per year on dredging operations that did not work (Lenton and Walkuski 2009:18) and were at the time alarmed at the prospect of having to dig a new lake. The city government thus asked for a study from the I.I.S.W.C., who found that most of the silt stemmed from an eroded catchment area in the hills to the north-east of the city. In this catchment area was a small village that, in the 1970s, was defined, as many villages were, as both poor and "underdeveloped". As in Rani Mājri, this village's inhabitants heavily relied on common forest land for agriculture and grazing to buffer the continuously impending threat of food shortage. The watershed management project that I.I.S.W.C. developed in this village became a success story that, despite its many difficulties and the disputes caused by caste and community disagreements, managed to turn underdevelopment to "progress", establishing a firm relationship between the ecological and socio-economic well-being of the hills and the plains.

The next life-changing technology introduced was fertilisers and pesticides (*khetī kī davāī*; lit.: "medicine for the fields"). The Green Revolution, to be unfairly concise, was introduced in North Indian hill agriculture in the 1960s and 1970s through higher-yielding varieties of grains. After a series of bad harvests in the 1960s, the central government abandoned its strategy of land reform and adopted the technological and commercialised Agricultural development of high-yielding varieties of wheat, developed in Mexico by the agronomist Norman Borlaug, who went on to win the Nobel Prize. Borlaug was, upon his invention, subsequently invited to India in 1963, where the new variety of grain was met with enthusiastic optimism. The United States was, along with the Ford and Rockefeller Foundations, "a major source of funding for carrying the Green Revolution to Asia", and the head of the United States Agency for International Development, William S. Gaud, was in 1968 quoted saying, "Assuming reasonable monsoons, India clearly will be self-sufficient in food grains in three or four years – no question about it" (Miller 1977). In addition to changing who was able to shift to the more capital-intensive varieties, the Green Revolution also changed relations be-tween castes, lineages and families (Berreman 1978; Beck 1995).

In Rani Mājri, chemical pesticides and fertilisers, Agricultural credit, modern irrigation systems and modern machinery, such as tractors, made their debut as late as the early 1980s. According to the farmers, they first be-gan to apply fertilisers and pesticides around 1981, and, for almost 20 years (up until 1998, to be exact), they would apply only a little, a tad here and there. Landowners saw great advances in yields during this period, accompanied by a rapid growth in economic wealth, as surpluses could now be sold by the large landowners, who were mainly from the Rajput caste. In 1982 a public bus transport service started to run from a village a little further down the hill. Although the bus had a reputation for being unreliable, and commuting to the larger towns became easier, especially for educational purposes. The middle and secondary schools in the vicinity had, up until the 1980s, been located too far away for many of the schoolchildren, especially girls, to at-tend. With this development, the 1980s allowed women in the village born in the late 1970s to complete 10 years of schooling. In 1983, another large intervention in the landscape of Rani Mājri occurred, when parts of the *kuhl* (the part leading through the village centre down into the fields) were made "semi-firm" (semi-*pakkā*) as a result of a village council (*Panchayat*[6]) initiative. As the old *kuhl* was made of stone and earth, it was a fragile open water channel, suffering from severe seepage, frequently running off into the river valley unless constantly maintained. With a cemented base, more water could be led to the village, and the fields could receive more water, making them more productive.

Around the same time (the exact year is unknown to me) a sub-surface water tank was constructed with funding from the local government, just above the waterfall. The tank was a closed, sub-surface construction and was intended to retain some water for dry-season use, and pipe some in as drinking water to certain landowning households (the design and implica-tions of which I will return to shortly).

As we move into the 1990s, media and communication accessibility now developed at great speed. In 1991, the first television was bought by a villager, and news and entertainment could be accessed in image and sound, making

Figure 2.2 The cemented main *kuhl* and the village above.
Source: Author.
Image showing a cemented water channel on the fields below a village.

the radio unpopular and left unplugged on the shelves. In 1992, further new and improved crop varieties were introduced to the farmers, and, again, there were notable increases in yield and profits (Yadav et al. 2008:16). In 1993, the Haryana State Government attempted to mitigate the underdeveloped condition of the region by appointing an independent governmental implementation wing, the Shivalik Development Agency (2017, 2020), as well as numerous national and regional schemes to boost the region out of poverty.

By 2002, village children no longer had to travel by bus or foot to receive an education after the age of 11 as a middle school (grades seventh to eighth) opened in the village itself, with government teachers traveling via bus and auto-rickshaw from urban cities and towns, on rotational duty.[7] A government-funded day care (*anganwadi*) for small children opened in 2003, and the first mobile phone was bought by a villager around 2006.

Many villages in the region had undergone projects to ensure ecological sustainability through soil, water and forest conservation during the previous decade, but Rani Mājri was not part of any large-scale, partly internationally funded projects until the I.I.S.W.C. began their watershed management project there in 2007. Technically, a "watershed" is the geographical area that rainwater flows through and drains into – a common body of water. These water management structures are meant to aid both the environment and economic development as improvements in the local irrigation channels (*kuhl*) both reduce erosion and enhance Agricultural production through a more reliable irrigation system. Whilst un-irrigated land in these hills allows for only one harvest a year (during the monsoon season), irrigated fields allow for a second, winter harvest. Five villages and their respective *kuhl* systems would be covered by the watershed management project. I.I.S.W.C.'s strategy of soil and water conservation in these severely eroded areas is maintained through the construction of "watershed harvesting techniques", which capture run-off water in the respective village or town by constructing small check dams[8] and solidify existing *kuhl* systems. The I.I.S.W.C. also initiated the planting of certain trees (such as the Kheel) and grasses (like the *Elopsis Binata*, locally known as *Bhabbar*). This vegetation has complex and deep root-systems well suited to hold soil in place in the steep hills. Other soil improvement measures were also taken, such as increasing organic farming, introducing bio-fertilisers, facilitating bio-fencing against wild animals and providing advice on crop-rotation cycles and choice of seeds as well as fostering a social entrepreneurial side by offering self-help groups for women. I will get back to several of the implications of the project later on, but, in general terms, it was regarded as quite successful by the team of scientists as well as by the larger landowners in Rani Mājri. The farmers had willingly embraced most of the initiatives from and advice given by the scientists, and the project managers enjoyed high status among those farmers directly impacted by the project.

The protective measures of forest management in the area also intensified when, in 2009, the Haryana Government declared the forest – an island of green in a sea of fields – below Rani Mājri a 760-hectare Wildlife Sanctuary (Dash 2009). Simultaneously, the village and the surrounding area was enveloped into a surrounding "Eco-Sensitive Zone". These "Eco-Sensitive Zones", which I return to in Chapter 3, are geographical zones where any polluting or highly polluting industry, wood-based industries or mining activities are prohibited (Haryana Forest Department 2016).

Contemporary Rani Mājri

In 2013, the fields of Rani Mājri looked like an extensive and intricate mosaic of colour and variety. The village and its surrounding hills and forests, however, appeared a wan and worn version of the bountiful forest and the rich wildlife described by the colonial gazetteers a century before, and trees like *Khair*, were few and far between so that lopping for fodder or collecting branches for firewood became a hazardous and time-consuming practice. Some bushes seem to thrive in the dry soil, especially the wild lantana bush. Its orange flowers are beautiful, but although they are pleasing to the eye, the bush has a recent and troubling presence in these hills. Farmers would tell me that, before the 1990s, it was barely present, but it had since expanded into their farmlands as an aggressive weed. The large and once vigorous river that had to be traversed by foot or elephants in the 20th century was reduced to an intermittent river with the completion of the highly controversial Kaushalya Dam (Sharma 2014; Seghal 2015), and a new wildlife centre was established in the forest reserve below the village to conserve animal habitats within the reserve.

The recently established wildlife centre was not received with any enthusiasm. The villagers were, instead, concerned about what to do with an eventual increase of wild animals that would eat their crops as few could afford the fencing required to keep them out. Many wild animals still found in the forest and in the hills – the large antelope (*nīl-gāy*) and the antelope stag (*bārahsiṅgā*), for example – are a nuisance as they wander into the fields at night to graze on the growing crops. Wild boars dig up newly planted roots, monkeys steal wheat and maize left to dry on rooftops, and the black leopard (*čīta-baghīrā*[9]) roams the hills to feast on village goats. Also moving about in the surrounding thicket were scorpions, lizards and venomous snakes of every size, from large and intimidating pythons and boas to smaller, but still venomous, kraits and vipers.

In Chapter 1, we saw that poverty hits families across caste identities as Lohar, Rajput and S.C. households would at times lack the means to provide sufficient nutrition, clothing and higher education for most or all members of their households. Still, there were decisive caste and class differences, and the largest – how able a household was to deal with periods of want – seemed to be related to the *kuhl* and thus to landholding. As such, the entwinement of caste status, land entitlement and economic status manifested in the right to utilise *kuhl* water for irrigation and household purposes.

Water was shared amongst the landholders according to strict rules for diverting the water between them at set times in a system maintained and coordinated by them at regular meetings. This distribution happened as per a "*barabandi*" system (Yadav et al. 2008:7), a part of an older form of administration of water resources. During my stay in the village, I did not register anyone talking about this system, neither did it occur to me that I should have asked anyone about it, but it seems to relate to a system of water allotments. The village had originally been divided in to eleven *sāmīs* (translating as ferule, rich arable land). One *sāmī* would be given twelve hours of irrigation water (during the rainy season it would be six hours' worth), distributed equally among these eleven *sāmīs*. Then, the extended families within the original 11 *sāmīs* distribute the same water among themselves, and so on (Yadav et al. 2008:7). In practice, this allocation system would happen by manually diverting the stream of the *kuhl* by filling openings with stone, earth and debris. The operation must happen quickly and precisely. In the growing seasons, mistakes or complications in adjusting the water quantity could mean the success or failure of a crop. As one can see

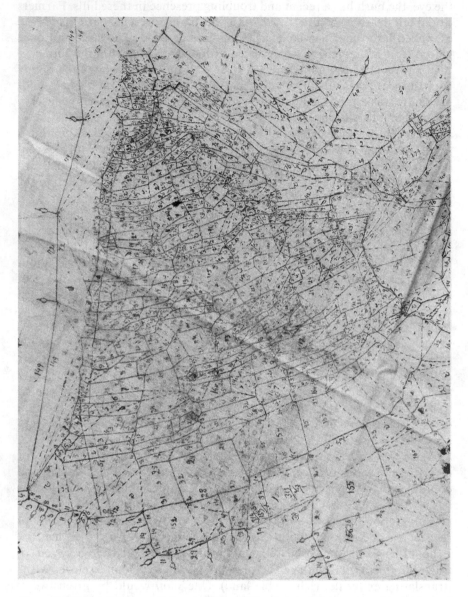

Figure 2.3 Map showing all the tiny plots of land, hand-drawn on a large piece of
cloth. The cloth measures approximately 1 m² and was first drawn in the
1960s.

Source: Author.

Map showing all the tiny plots of land, hand-drawn on a large piece of cloth.

from the map below, the system is rather intricate, and the rules of allocation have only grown more complex over time.

Those farmers with rights to *kuhl*-irrigated land have always had a greater
quantity and a more stable supply of water. The later refurbishment of the

kuhl provides them with more of a buffer in the sense that the tanks in the irrigation system store the water and provide gradual distribution. In many ways, the development and improvement of the *kuhl* system and agriculture have benefited the larger, male, higher-caste landholders in many ways. Owning a minor plot, however poorly irrigated by being peripheral to the trajectory of the kuhl, also appeared to entitle a family the privilege of utilising *kuhl* water for drinking, cooking, laundry, etc. The I.I.S.W.C. watershed management project, especially, appears to have brought social, symbolic and financial prosperity to one segment of the population – high-caste landholders. Landholding families with a *kuhl* irrigation entitlement can grow food for their own consumption and sale on more fertile and productive land, choosing expert-recommended varieties of crops that are more sensitive to water-distribution and grown more effectively, thereby providing yield in both the winter growing season (*rabi*) and during the monsoon growing season (*kharīf*). From selling the surplus of the harvest or from the income from white- or blue-collar labour, a household might allot itself money to buy vegetables from the vegetable truck arriving every fortnight, purchase milk when the milk cow or buffalo was pregnant or sick, buy a fridge to keep food fresh for longer or pay for both its sons and its daughters to continue their education in college. They can afford to pay a decent dowry, too, so their cherished daughters can marry into respectable homes as adults, between the ages of 18 and 22, not at 12, in which case they would become mothers at 15, as in previous generations. In fact, many state and privately driven initiatives and projects done in the name of climate change adaptation or mitigation in the region, such as building dams, highways and water harvesting structures; restricting the use of the forests; enhancing efficient Agricultural methods; and increasing the level of technical expertise and the right kind of awareness through education have created alternative paths to progress that have largely benefited the farmers positioned to capitalise on these changes.

The "landless", however, become disconnected from both development itself and the ability to actively shape it.

Water rights

Southward from the village centre, another earthen *kuhl* ran along the path towards the school building, the S.C. hamlet and eventually Khot. Below, irrigated fields fan out towards the west. Towards the east lie the steep hills, and a flight of stairs leading up to the small temple devoted to the village guardian deity, Kheṛa Baba (see Chapter 4). The *kuhl* would seldom bring water towards the southern fields and the S.C. hamlet, however, only in the height of the monsoon season did the open trench flow with water. In the monsoon of 2013, it did not seem to amount to much. Perhaps it was the season, but the S.C. population seemed to think it was related to the practices of the Rajput and Lohars in the main village. Incidentally, a small, surfaced pipe carried drinking water southward along the same pathway to the smaller Rajput village of Khot - straight past the S.C. hamlet.

The S.C. settlement in Rani Mājri was a small collection of houses, just below the path that connected the main village with the neighbouring village of Khot. There were only eight households here, and no real village centre. The houses sat in a row along a small path that was part stone, part dust. On the surface these looked newer and "more solid" than many Rajput and Lohar houses as they were relatively newly constructed from cement – simple one-story buildings with one to three rooms, painted in a light yellow colour. The construction of these *pakkā* houses was, I was told by the S.Cs, "given by the Government", most likely as a part of the Indira Awas Yojana (I.A.Y.), a national programme of rural housing subsidies implemented through the Rural Development Department (Drèze and Khera 2010:56).

All S.C. households had black water tanks outdoors to store water for their separate washing nooks. Upon entering them, they were quite like any of the smaller houses of the main village. The same wedding photos, deities and movie star posters alongside the cracks of paint on the wall, chest of tin to keep their things, humming old fridge and boxed TV-set, naked lightbulb hanging from the roof and household shrine – albeit with a few different deities. The S.C. hamlet had, however, hooked into the electrical grid much later than the main village (in the early 1990s) and had no paved road access. As their marginal land was unirrigated by the *kuhl*, they were also excluded from the I.I.S.W.C. watershed-management project, and they relied completely on government well water, which was groundwater pumped up to the village in small pipes for a few hours a day for a fee – which, in 2017, was at approximately Rs. 2 per kilolitre of water in a rural, domestic household (including a 25% bill fee) (Haryana Public Health Engineering Department 2017). Seeing as that number is affected by the everyday practice of bribery in the form of exchanging bottles of alcohol, favours and money, the actual price depends on who you are and what socio-political leverage your family holds, and most people I talked to found government water to be expensive, impractical and unreliable. It could not be used for irrigation purposes as it was accessible only for a few hours in the morning, just enough to refill buckets and water-containers for drinking, laundry and personal hygiene. It was, however, appreciated as being filtered and clear. Drinking the water flowing through the kuhl and the small perennial streams during monsoon was hazardous (everyone knew there were "germs" that could make you ill in the water, from defecating livestock and settlements above), and unpleasant as it would be heavily silted and murky, so it was directed to constructed ponds or groves, where only livestock could use it to quench their thirst.

The S.Cs had no entitlement to the *kuhl* as – the S.Cs told me – they were "landless". However, landless appears in quotation marks here. The S.Cs did, in fact, have land that they were entitled to – the presence of their field deity's shrine gave testimony to Avani's observation - but in practice, the land was fallow. The S.Cs had access to a marginal area of *rain-fed* land. This meant that the land lay fallow during the growing season of *rabi,* as the winter rains were insufficient for growing the modern, more water-needy variety of seeds (Mehta 2011). Without irrigation to stagger overflow and save the surplus water, rain-fed land is also more vulnerable to variations in the heavy south-western monsoon during *kharif.* Investing in fertilisers and pesticides came at a high cost, which meant what they could grow was susceptible to disease and pests. With little fertile land, there would also be less animal fodder, which meant that owning cows or buffaloes became too expensive for the S.Cs. With no bulls to till the land, the work needed to prepare it for sowing would require much time and labour. The S.Cs would thus need to spend more money on food, at least without the ration cards that the Below Poverty Line (BPL) status gave them. Working at the landholders' farms during harvests would be important for payment in grains such as millet, maize and wheat. For the S.Cs, rainfed land was as good as no land, and their "landlessness" as factual as it was misleading.

The issue of landlessness was sensitive, and S.Cs never admitted to me that the land was theirs to grow. I have reasons to believe this was based on concerns that I would report their assets to the government, who could then categorise them above the poverty line, thereby depriving them of BPL status, as mentioned in Chapter 1. Developed from a measure of hypothetical caloric intake at a certain level of income, the definition of households above or below the regional BPL has become an intricate practice, with parameters varying from state to state. How the line is fixed is vigorously debated by both social scientists and economists,[10] but in Haryana, BPL status is set on the basis of a highly diversified list of property and at an income benchmark of Rs. 27 ($0.45) a day for rural areas and Rs. 33 ($0.55) a day in urban ones.[11] This is lower than the World Bank's official poverty line (which was at $1.25 in 2011–2012[12]) or what Drèze and Sen (2020) call a "destitution line". Then, an additional 14 parameters would indicate automatic exclusion, such as if the household had any member serving as a government employee, if they paid income or professional taxes, if they had three or more *pakkā* (cement or brick constructed) rooms, if they owned a refrigerator or a landline phone, or if they had over 2.5 acres of irrigated land. There are also automatic inclusion categories, such as those without shelter, scavengers and the destitute living on alms, as well as deprivation criteria, such as households with no literate adults over the age of 25 or female-headed households with no male member between the ages of 18 and 59 (Haryana Rural Development Department BPL Note, undated). To qualify for ration cards in Rani Mājri, for example, the household had to be BPL, a status many had lost based on a census conducted in the village around 2011–2012 (Planning Commission of India 2017); all the S.C. households held ration cards, as did a few Rajput and Lohar families. After that survey, most of these households lost their eligibility for these cards as the S.Cs had been provided certain "baseline assets" through various schemes like the "Swacch Bharat Abhiyan", officially launched by the Government of India in "entire rural India" in 1999 (Ministry of Drinking Water and Sanitation 2016), which provided funding for the construction of "latrines", as well as I.A.Y. (see above), which allowed for the construction of cement houses. These ration cards were sorely missed, especially by the poorest families, as they gave the right to free or strongly discounted prices on certain staple foods and household items.

Consequently, the S.C. population were "landless" in the sense that they only had access to rain-fed, hard-to-reach plots; thus, their dependency on outside employment. Many of the women were busy herding their goats, working for the government as street-sweepers or washing in the primary school (positions that would pay Rs. 500 a month); young boys would quit school at an early age to work in the informal work sector, with irregular day-to-day labour, no pension and no other workers' rights. With marginal incomes, an S.C. family would often send male youth to do factory work before they could complete the 10th standard in order to contribute to the economy, and girls were generally married off earlier than in the main village, where the age of 18 seemed to be a normal age for marriage. Marriages here also often tended to be "joint" as brothers could marry in one ceremony so as to save expenses.

Unirrigated development

The S.Cs' lack of access to *kuhl* water was not an issue any villagers in Rani Mājri wished to discuss further with me for many different reasons (as political power was in the hands of Rajputs, there was nothing to gain from a falling out with them). But on some occasions, the matter did arise. I first began making inquiries about the *kuhl* water in the hot and dry month of April, when there was much dismay all around, since the Government well water from Bilaspur had "gone" again. On the third day with no one getting any water from their taps, their tanks were running dry. Waiting for someone to repair the alleged broken pump, the situation was particularly acute in the S.C. hamlet. Upon visiting a woman there, I told her I had seen people from Khot use water from the pipe that looked like it was coming from the Rani Mājri *kuhl* and asked why they had not done the same. The S.C. woman shrugged and told me that that village water was no good for them. She did not know what was wrong with that water, but it was something. Their hair got all sticky when they bathed in it, and they got ill if they drank from it, so they would never use it. She might not have known how little I knew of local caste regulations at the time, but she knew very well whom I lived with. Her answer, that the S.Cs found the water running in the *kuhl* unsuitable for their needs, puzzled me.

A few weeks on, I went with my son for an evening stroll and was happy to be invited into an S.C. house I had not visited before. I was called in by a group of males, where one visiting uncle, probably a bit tipsy, began to loudly voice his opinions on the Rajputs. The other men grew increasingly uncomfortable with his replies, especially when I asked about the *kuhl* water. "There was water in the *kuhl* before", one of the men answered initially, "but that was a long time ago. Perhaps a hundred years back". "And now it is cut because we are Harijan", the uncle interrupted rather aggressively, "and the people with responsibility are doing nothing!" The men were so uncomfortable at this point that I decided to change the subject. Later, I asked Prakash about the S.C. access to *kuhl* water and if it had anything to do with their caste, but he denied any correlation with the lack of water, saying only that "they have no land, and so they have no water from the *kuhl*", and if they were so eager to get water, I could tell them that I would "bring some from Norway". I did read his comment not as unwillingness to share with the S.C. per se but as the idea that, whatever little there was, he was entitled to it, and they were not. That their caste status might have influenced their lack of land to begin with was not seen as an issue.

The purely class-based division veiling how caste intertwines with water rights in its economic guise can thus be argued to predominantly favour not only the largest landholders but the Rajput caste in particular. This is relevant as, in Rani Mājri, households do organise themselves partially according to something that resembles a *jajmāni* structure of village order. A *jajmāni* system refers to how the landowning castes produce and distribute food in a shared village economy. In early studies of the *jajmāni* system, the landholding castes were thought to produce and distribute food to the servant castes of the village (the barber, the carpenter, the smith, etc.) in the form of a closed village economy (Raheja 1988a,b). As an example, from Rani Mājri, the S.Cs that would assist the relatively large landholders during the most intensive harvesting periods were paid in produce. Produce was also the custom payment for leasing farming machinery, such as tractors for threshing, as per the *jajmāni* system.

That trading of service for produce might have appeared a mutual or symmetrical system, indicating grain exchanged for services in a tit-for-tat manner, apparently beneficial to everyone involved (Mines 2005:60). Later analyses of Indian villages came to see those relations as not mutual at all but as largely coerced and connected to control over the food supply, with the landowners using their power as "dominant castes" (ibid.:60). Nor is a *jajmāni* relation, where it exists, enclosed within the village; rather, it is and probably always has been part of wider relations in political-economic systems "of patronage, kingship and trade" (Mines 2005:61).

The Rani Mājri of 2013 was a village that had undergone great changes – all remembered by a living generation. From once having no access to electricity, education, media, entertainment or piped water, by then all households had televisions, nearly all had piped-water, everyone born after 1980 had an education and most children of both genders born after the turn of the millenium had completed 8th grade which completes compulsory right to education in India. The youth of both genders were also increasingly attending and completing twelve years of education ("plus-2"), the higher secondary school level that prepared them for college. By 2013, a handful of males and one woman had travelled to larger cities in the vicinity to pursue higher education in government colleges. Khot, it was rumoured, would soon have a paved road too.

There is no doubt that various rulers – be they Sikh, Mughal or British – have consciously shaped and reshaped relations between those who have and those who have not. Clearly, "development" – as well as the managing of water, soil and forest in Rani Mājri – is closely related to maintaining beneficial relations with the State Government. The next chapter will focus on those practices of the government and how it affects the perceptions of the "environment" in Rani Mājri.

Notes

1 When Avani referred to the S.Cs of the village as "Harijan", she came across as trying to be polite by not calling them by their caste name, Chāmar, which was thought to be extremely derogatory and was only once whispered to me by a child, then confirmed by adult nods. Harijan was the term chosen by Gandhi to refer to the S.C./Dalit/Untouchable population of India; it is used mostly by the higher castes, and sometimes by the lower castes themselves, but is publicly seen by many as offensive and equally derogatory.
2 There are four possible Navarātras, but spring and autumn are the most popular in the village as they take place after two important sowing periods.
3 Kathleen Erndl (1998), who studied devī possessions in Haryana and Punjab, notes that Durgā has a widespread appeal and that in the region of North West India, the Goddess is approached in some form or the other by both high and low castes, urban and rural populations, and Sikh and Hindu households (Erndl 1998).
4 I have been notified that 'atma shanti' is mentioned in Hans Hendriksens 'Himachali Studies' Vol. 1 (of 3) from 1976, but I have not been able to retrieve this document. I will thus refrain from attempting to translate this concept.
5 A pandit is a Brahmin learned in the Vedas, the sacred scriptures of Hinduism, and a Brahmin is the highest/purest Hindu caste.
6 A Panchayat (Pancāyat) is a local village governing system that gradually took over from the hereditary Lambardar tradition. In the Panchayat, a representative leader titled 'Sarpanch' is elected every fifth year.
7 The Haryana State Government operates a Transfer Policy, which implies that government teachers can be shifted to vacancies around their district as part of their duties, with a normal duration of each stint being 5 years (Haryana School Education Department 2016).
8 A check dam is a small dam constructed across a drainage ditch or channel to lower the speed of water flow. This allows sediments to settle and groundwater to recharge (Stauffer et al. 2017).
9 They called it cīta (cheeta) and baghīrā (bagheera) interchangeably, even though it is not the same animal. A bagheera is a black leopard and a cheetah is a spotted type of puma.
10 See economist Drèze and Khera (2010), Drèze and Sen (2013) and Guruswamy and Abraham (2006).
11 This has been changed several times – latest in 2019 – when, according to the Times of India, the maximum income for BPL status was set at Rs. 15,000 a month (Times of India 2019).
12 Adjusted to $1.90 in 2015 (World Bank 2015).

References

Ambala Imperial Gazetteer
1909   Panjab University, Chandigarh.

Baker, M.
2005   Kuhls of Kangra: Community-Managed Irrigation in the Western Himalaya. Seattle and London: University of Washington Press.

Baviskar, A.
2001   Forest Management as Political Practice: Indian Experiences with the Accommodation of Multiple Interests. *International Journal of Agricultural Resources, Governance and Ecology* 1 (3/4): 243–263.

Beck, T.
1995   The Green Revolution and Poverty in India. A Case Study of West Bengal. *Applied Geography* 15(2): 161–181.

Berreman, G.D.
1978   Ecology, Demography and Domestic Strategies in the Western Himalayas. *Journal of Anthropological Research* 34(3): 326–368.

Cunningham, F.
1883   Ambala District Gazetteer. Haryana: Revenue Department.

Dash, D.K.
2009   7 Forests Including Sultanpur Declared Eco-Sensitive Zones. In: *Times of India*. http://epaper.timesofindia.com/Default/Layout/Includes/TOINEW/ArtWin. asp?Source=Page&Skin=TOINEW&BaseHref=CAP%2F2009%2F07%2F06&-ViewMode=HTML&GZ=T&PageLabel=4&EntityId=Ar00401&AppName=1, accessed November 19, 2015.

Drèze, J. and Khera, R.
2010   The BPL Census and a Possible Alternative. *Economic and Political Weekly* 45(9): 54–63.

Drèze, J. and Sen, A.
2013   An Uncertain Glory: India and its Contradictions. Princeton; Oxford: Princeton University Press. doi:10.2307/j.ctt32bcbm

Erndl, K.M.
1998   Seranvali: The Mother Who Possesses. In *Devi: Goddesses of India*. John Stratton Hawley and Donna Marie Wulff, eds. Pp. 173–194. New Delhi: Motilal Banarsidass Publishers.

Gazetteer of the Simla District
1888   Delhi: Hillman Publishing House.

Grewal, K.
2004   British Paramountcy and Minority Administration: A Case Study of Patiala (1900–1910). *Proceedings of the Indian History Congress* 65: 646–656. Retrieved August 21, 2020, from http://www.jstor.org/stable/44144779

Guruswamy, M. and Abraham, R.J.
2006   Redefining Poverty: A New Poverty Line for a New India. *Economic and Political Weekly* 41(25): 2534–2541.

Haryana Forest Department
2016   Protected Area of Haryana. *Government of Haryana*. http://www.haryana-forest.gov.in/protect.aspx, accessed November 22, 2016.

Haryana Public Health Engineering Department

2017   Rural Billing. *Government of Haryana.* https://biswas.phedharyana.gov.in/ WriteReadData/Notice/CurrentTariff.pdf, accessed July 6, 2020.

Haryana School Education Department

2016   Teachers Transfer Policy 2016. Government of Haryana. http://harprathmik.gov.in/pdf/circullers/TransferPolicy2016.pdf, accessed December 2, 2017.

Lenton, R. and Walkuski, C.

2009   Integrated Water Resources Management in Practice: Better Water Management for Development. Mike Muller, ed. UK and USA: Earthscan.

Miller, F.C.

1977   Knowledge and Power: Anthropology, Policy Research, and the Green Revolution. *American Ethnologist.* 4(1), Human Ecology (Feb., 1977), pp. 190–198.

Mines, D.P.

2005   Fierce Gods: Inequality, Ritual, and the Politics of Dignity in a South Indian Village. Bloomington and Indianapolis: Indiana University Press.

Ministry of Drinking Water and Sanitation

2016   ABOUT NBA. Swachh Bharat Mission – Gramin. *Government of India.* http://tsc.gov.in/TSC/NBA/AboutNBA.aspx, accessed November 22, 2016.Pemble, J.

2009   Forgetting and Remembering Britain's Gurkha War. *Asian Affairs* 40(3): 361–376.

Raheja, G.G.

1988a   India: Caste, Kingship, and Dominance Reconsidered. *Annual Review of Anthropology* 17(1): 497–522.

1988b   The Poison in the Gift: Ritual, Prestation, and the Dominant Caste in a North Indian Village. Chicago: University of Chicago Press.

Rajadhyaksha, A.

2017   Kingdoms of South Asia – Minor Indian Kingdom of the Jat Sikhs. Jat Sikh Minor Kingdoms. http://www.historyfiles.co.uk/KingListsFarEast/India-JatSikhMinorKings.htm, accessed January 18, 2017.

Seghal, M.

2015   Scam Worth Rs 217 Crore behind "Failed" Kaushalya Dam. *Daily Mail Online.* https://www.dailymail.co.uk/indiahome/indianews/article-3016245/Scam-worth-Rs-217-crore-failed-Kaushalya-Dam-says-CAG.html, accessed January 19, 2017.

Sharma, P.

2014   CM Orders Vigilance Probe into Kaushalya Dam Work. *Tribune India News Service.* http://www.tribuneindia.com/news/haryana/cm-orders-vigilance-probe-into-kaushalya-dam-work/22153.html, accessed January 19, 2017.

Shivalik Development Agency

2017   Demographic Profile. *Government of Haryana.* http://www.sda.gov.in/ Page.aspx?n=137, accessed April 4, 2017.

2020   Shivalik Hills Area Description. *Government of Haryana.* https://sda.gov. in/en/our-profile/area Accessed June 2020

Stauffer, B., Carle, N. and Spuhler, D.

2017   Check Dams & Gully Plugs | SSWM. http://www.sswm.info/content/ check-dams-gully- plugs, accessed April 5, 2017.

Times of India

2019   Minimum income limit up for BPL cards. *Tribune News Service.* https://
www.tribuneindia.com/news/archive/haryana/minimum-income-limit-up-for-
bpl-cards-820775, accessed June 2, 2020.

World Bank

2015   Country Dashboard. *The World Bank: India.* http://databank.worldbank.
org/data/Views/Reports/ReportWidgetCustom.aspx?Report_Name=Coun-
try_chart1_June4&Id=5ee1de8357&tb=y&dd=n&pr=n&export=y&xlbl=y&ylb-
l=y&legend=y&wd=430&ht=380&isportal=y&inf=n&exptypes=Excel&coun-
try=IND&series=SI.POV.NOP1,SI.POV.DDAY, accessed September 14, 2015.

Yadav, R.P., Singh, P., Arya, S.L., Bhatt, V.K. and Sharma, P.

2008   Detailed Project Report: For Implementation under NWDPRA Scheme.
*Central Soil & Water Conservation Research and Training Institute, Research Cen-
tre, Chandigarh.* New Delhi: Ministry of Agriculture, Govt. of India.

3   Governing awareness

Hemant

Hemant is the season of "late autumn", consisting of the two lunar months of Kārttik (mid-October to mid-November) and Mārgaśirṣa (mid-November to mid-December).

The nights in late autumn are increasingly nippy, but the days are warm. Butterflies enjoy the heat of the sun and flutter in between the orange and pink flowers; the birds tweet, chirp and chitter. There has been no rain for weeks, and the foliage is increasingly withering as it gathers dust blowing up from the dry ground. The water in the kuhl has lost its ferocity now, and the stream is unable to carry waste downhill. Plastic and paper residue gathers in the small nooks and twists of the timid stream that flows gently onto a yellow world of maize at full height and rice turning ochre. It might seem quiet and peaceful, but the village simmers with activity from early dawn until late dusk. The harvest of the monsoon crops happens alongside the sowing of winter crops, so young children come home after school and run down to the fields with tea or water, while older children work with their parents to harvest, rinse, bundle, sort and carry. The double workload requires larger landowners to recruit able hands outside their lineage too, most often S.C. labourers.

As the maize is harvested, mustard (sarsoṁ) and chickpeas are planted. Then, the valuable ginger is harvested and prepared for drying. It is tedious work. The roots are rinsed, then cut into small pieces, which are spread out on almost every roof in the village. In the dry weather, it takes the roots a few weeks to turn white and light, then they're able to be sold as dried ginger (soṅṭh). The rice is also harvested now and stored indoors in various metal crates. The grass is dried and stored together with maize stems for additional dry fodder throughout the winter. The tomatoes ripen in bulk and are carried from the fields in blue plastic crates. The fresh coriander is also harvested and rinsed, carefully bundled together; surplus vegetables, if any, are sold.

As the season moves towards its end, the sky turns dense with the cool mist of the night. The white veil blends with grey dust and smoke from exhaust and farmers' fires as they burn the residue from the last monsoon harvests on the fields. The smog is pressed down to the ground by thermal inversion on the plains. In Chandigarh, as in many other large cities on the plains, the air becomes thick with particles, and the sun changes colour from bright yellow to dusty red. Winter is on its way.

To prepare, women from all households venture into the adjacent village forest frequently to collect firewood, fencing and leaves for fodder and rituals. These trips take time, hours, in fact, as the hillside holds scant trees that can be cut, and the women must walk and climb long distances to find suitable branches. Mostly, the women did not want me to come along; it was dangerous, they said, as they ventured off, sometimes alone, sometimes in twos. One day, after I assured her I had climbed mountains before, Avani let me tag along. "Dress in old clothes", she told me, "and swap your sandals for shoes, and meet me tomorrow morning after you have eaten!"

Early next morning, I met Avani, dressed in her husband's old jacket and trousers to cover her own śalwār-qamīz suit, and we began our ascent. We followed a path at first, through the thick and thorny undergrowth, leading us past a tree. It was decorated with faded threads, which must have been bright red or orange at some point, vaguely glimmering with silver bits of plastic tied to them. This was Chandi Devī's shrine, Avani said. She told me she would always take care to greet the devi when she passed the site, so we nodded towards the shrine, and lightly touch our foreheads and our chests with the fingertips of our right hands.

To find trees suitable for lopping or firewood, or the right leaves for ritual purposes, one must venture off the paths, and here, the hills were difficult, even treacherous, to climb. In some places, we had to hold on to roots and branches exposed by the eroded hillsides out of fear of going the way of the loose stones on the sandy slope: down the ravine to the riverbed. "These are not like the hills at home", I said, obviously nervous. "I never worry", Avani replied.

Along our way, we came across several useful kinds of twigs and leaves. Jiggṛi and Butti was collected for firewood, Bijūl and Bamboo for fodder, and Sarali for making ropes, and the Khajoor (the date palm) would be lopped for its leaves at the very top, which were ideal for making brooms.[1] The thorny bush, Kaṇte, was popular for fences as their sharp thorns and spikes are efficient at keeping animals away from the fields, but those were often left to be harvested by the men as they tore badly at hands and clothes. The real prize, Avani told me, was the Khair. "It is the best firewood", she said, "and it gives a high price!" But unless you had one on your own land, which Avani did not, Khair could not be logged. Sticks and branches that had been broken off by storms or monkeys, however, were for anyone's taking. With a hasty step, Avani went into thorny bushes, climbed steep sandy ridges and slid down again with a particularly useful specimen.

The Neem, Mango and Ber are also valued trees that grow close to the village. Neem provides daily-use dental sticks and herbal medicine, and Mango and Ber provide the sweet fruits of the mango and jujube (Indian plum), respectively. The Kraḷti[2] tree, one particularly good specimen of which grew in the centre of the S.C. hamlet, was valued for its all-round use as the large pink and white flowers were used in the yogurt sauce of raita, the buds as a vegetable in dishes (sabzī) and the leaves as fodder.

On our way back, we came across the Bael – a tree cherished for its tri-foliate leaves used in the worship of Shiva – and the wild Amla (wild gooseberry). In late autumn, the berries are still unripe, small and green. Avani showed me how to climb the small, bush-like trees to bring a few home for the children – they were fun, she said – and made me chew one; it was so bitter and tangy that it made my eyes water.

Closing in on the village, we paused to take in the sight of the fields of Rani Mājri below us. The winter wheat was maturing in the fields and dominated the palette with its dark green hue, but the collage had significant patches of chick-peas in bright green, mustard blossoming in bright yellow, and garlic and onion with purple and white flowers. As she bundled the firewood to carry home, she was suddenly distracted by something across the hill. Following her gaze, I saw two goats enjoying a nibble of the thicket on the other side of the riverbed. "Oh no, they're mine!" she suddenly exclaimed, and, in the blink of an eye, she was on her feet and skidding downhill, sand and stones following her, whilst she turned back and told me not to follow her further than the path. I could only watch her run across the dried-out river and up the other side, where I heard her calling her goats over to the right side of the hill.

"That is not our forest", she said, sweat pouring down her forehead. I asked her whether it belonged to the village on the opposite ridge, but she shook her head. "No, it's government". If the villagers were to cut the trees there, she explained, or let their goats graze, the Forest Guard would cause problems.[3]

....

The I.I.S.W.C. is currently participating in the development of numerous soil- and water-conservation related projects in the Shivalik Hills. They do not act solely on behalf of the Indian state but also collaborate with large international actors, like the World Bank, the Swedish International Development Cooperation Agency (S.I.D.A.), the Danish International Development Agency (D.A.N.I.D.A.) and the International Centre for Integrated Mountain Development (I.C.I.M.O.D.), in close collaboration with the United Nations Intergovernmental Panel on Climate Change (I.P.C.C) amongst others. The documents and their intentions are many, and their ambitions are high. However, "There is so much lost in translation, from the ones in France or wherever, making the plans, to English, to Hindi, to the local dialects or to Panjabi", according to the scientists at I.I.S.W.C., who argue that it is unlikely for policies on climate change mitigation drafted in Europe to work seamlessly on the local level.

Mr R. C. Gupta noted the same problems with translating the ideals of conservation and development at a local level. Gupta was a middle-aged, retired state hydrologist who had co-founded the environmental, non-profit organisation the Society for Promotion and Conservation of Environment (S.P.A.C.E.)[4]. Working with the NGO and his later founded water-foundation, he had wished to facilitate the interaction between village and government post "mountain water system rehabilitation-projects", such as those carried out by I.I.S.W.C. During his time in the Department of Agriculture, he had been disturbed by what he saw as a lack of "interaction facilitation" between villagers and government officials. He observed that when, say, a watershed project finished, the government officers and associated researchers would withdraw completely, which made the project likely to fail in the future. But by educating the villagers, one could facilitate a transition, allowing them to carry on after withdrawal. As such, "the NGO arose out of pure necessity", Gupta told me. To Gupta, it was clear that, for the hill people to prosper in cohort with their environment, each one had to be taught to preserve and maintain their surroundings. But first, they had to be made aware of their role in the larger ecosystem. Understanding why one should protect the environment (especially the forest) and restore and repair the irrigation *kuhl* systems could only be achieved by transferring knowledge from those who knew to those who did not yet know – the environmentally unaware.

On global-local gaps and frictions

The need for raising awareness amongst rural hill villagers on issues of Sustainable Development thus seemed to be a recurring theme in most encounters between a "developing actor" and the "to-be-developed subject" in the Shivalik Hills in 2013. When government intervention failed to produce the wanted form of awareness, it seemed to be blamed on some sort of mistranslation, or that communication had not "reached" across a conceptual "gap" – black voids where information changes its expression, and the meaning gets lost, misunderstood or misinterpreted. This is often referred to as a problem of embedding the "global" into the "local" both by scientists and policymakers; thus, governance is often based on the necessity of bridging or closing that gap by transferring knowledge. Whilst the I.I.S.W.C. and the Agricultural department could provide the technicalities for conservation and development, individuals like Mr. Gupta, alongside numerous regional NGOs, strove to mend that "gap" between scientific policy and practical materialisation.

The idea of the relationship between the "global" and "local", however, is under heavy scrutiny, particularly by anthropologists. Through a close ethnography from the Meratus Dayak of the Indonesian rainforests, Tsing developed the metaphor of *friction* to explain that those gaps between the "global" and the "local" are not voids at all; rather, they are awkward zones of engagement – "real places into which powerful demarcations do not travel well" (Tsing 2005:175). Friction is neither necessarily beneficial nor destructive, but an indication of a sort of a creative process happening in the concrete meetings and engagements between actors in places where certain forms of knowledge and practices compete for attention. To further Tsing's argument, executing any project, scheme or plan – no matter how global and abstract it may seem – at some point always involves human interaction. In round-table discussions, conferences, offices, homes and over a meal, people exchange, alter, modify and adjust the content or practice of the plan or scheme to the context and issue at hand. Paying attention to these meeting points – these junctions of information travel – might tell us something important about how governing awareness happens.

Junctions

Very few people from the Indian Ministries or global organisations interact directly with the rural Shivalik Hills population, they mostly stop at the regional offices. The employees in the regional offices in Delhi, Mumbai or Bangalore seldom travel further than to their partners and colleagues in the district offices, and seldom talk to rural hill villagers at all. Nor do tourists, be they Indian or from elsewhere, on a trek or on their way to pray at sacred sites in the mountains. If they do make a stop in the foothills, they head for the more tourist-friendly eco-lodges of Shimla, Morni or Solan. Face-to-face interaction with the villagers for the purpose of educating, assisting or developing them is normally left for what we might call meso- or middle-level actors. For the region around Rani Mājri, this consisted of regular visits from the officers of the Forest Department, local government teachers, scientists of I.I.S.W.C. and Regional Rural Development officers. These junctions (of which I can outline only the few I encountered during my year-long stay) between the local and meso-level, mediated the ways in which information in the shape of scientific knowledge about global warming and environmental stress was transferred.

Junction 1: governing bodies

I remember well when I first learned about the village recycling system. It was in January 2013, and I had offered to help sweep the roof that had been the venue for a recent wedding celebration. Brushing the glimmering plastic, paper and various other sources of waste into a nice pile, I asked Bhagwati, Prakash's mother, what to do with it. She told me to just "throw it away!" (*phemk de!*) and brush it off the edge into the buffalo yard below, where they would later bring it with the manure to the fields. This was not my first experience with rural India, and I knew there was no other form of waste disposal in the village, so, doing as I was told, I swept the plastic waste down to the buffaloes. It soon turned out that there was always something to throw away. Empty plastic cans with fertiliser or pesticide, broken plastic chairs… all were thrown into the shrubs and thicket down the hill. Smaller items, most of which were candy wrappers or small plastic packets of shampoo or soap, were thrown into the *kuhl*, on the side of the road or in the forest. When I had to give up my attempts to use cloth diapers on my son due to his (and our) internal struggles to adapt to the innumerable amounts of new bacteria, I was told to throw his pre-perfumed disposable diapers downhill too. To many a raised eyebrow, I insisted they be kept in a bag for our trips to the city. I tried to explain that these diapers would pollute the landscape for a century at least before fully decomposing. Rekha, Prakash's youngest brother's wife, agreed that waste piling up was a problem but said that I should not worry about it. "When the monsoon rain arrives, it will take it all away", she explained, in what I perceived to be an "out of sight, out of mind" manner.

In June 2013, just before the monsoon season began, the Rural Department sent its head of office to hold "motivational talks" in select villages in the rural region. That year, Rani Mājri was one of them. The talk was part of the 1999 Clean India Mission (Swacch Bharat Abhiyan), revamped a decade later by Prime Minister Narendra Modi to "bring about an improvement in the general quality of life in the rural areas" (Ministry of Drinking Water and Sanitation 2016). The Government launched the campaign to achieve a clean (*nirmal*) state in India's villages by 2022 by adopting a "community led" and "people centred" strategy. When the day of the talk arrived, white plastic chairs and a teacher's desk were carried out from the primary school and placed on an open space at the end of the paved road. Another sunny and warm day, the desk was placed in the shade of a large tree for the government officers. Pausing their work, farmers – mostly Lohar and Rajput males – slowly arrived from their work in the fields. No one sat down on the chairs; they all just stood, waiting. They were impatient – there was other work to be done now, when the rain was to return to the village. Finally, the Government vehicles, shiny, white, off-road cars popular for these rural roads, arrived. A cameraman and a local journalist tailed the small cortege. The officer sat down behind the desk, briefly greeted the crowd, then began to speak in an authoritative and loud voice about the importance of keeping the village clean. He stressed how this could help to control the outbreak of expected water- and other vector-borne illnesses during the impending monsoon. He then proceeded to instruct the listeners on the importance of keeping their kitchens clean and gave a short description of how best to clean them (thinking, perhaps, that the men would bring this knowledge back to their wives, a couple of whom were looking at the spectacle from the safe distance of the roofs of their houses). Next, he added how people should wash their hands with soap after using the toilet. He completed his talk about the necessity of avoiding open defecation before he eventually opened the talk up for questions. A few hands rose. One male from a marginal Rajput landholder family was concerned about how to apply for money from the government to construct a latrine. Another question concerned the unknown outcome of a land-twist that had gone to court some while back and was yet to settle. The Q&A was a short affair, and, in the warm sun, the team quickly moved on to why the journalists had arrived – a public display. A few men from the S.C. hamlet who had kept to the background throughout the meeting now stepped forward with brooms, with which the officer and

Figure 3.1 The public display of the officer sweeping the street was meticulously
documented by the local press.

Source: Author.

Image showing a small group of people sweeping a street.

two other men from his crew proceeded to sweep a section of the street.
The journalists delighted, and, after a minute or so, the officer and his crew
departed. The sweeping display made the men grin awkwardly at each other
and the women on the rooftops shrug and shake their heads in light amusement before they all returned to work.

It was perceived as an honour for the village to be greeted with the officer's presence, but the way the visit was formal and distanced. No tea, no
sitting down together, no greeting or addressing any of the villagers directly
– this was a show put on for the journalists. The spectacle was received with
puzzled amusement by the crowd; to sweep public areas in the village was,
as everyone knew well, a defiling task to be handled by the S.C. members.

I later interviewed the head of Rural Development Department and his
team in person, months after the he had held the motivational talk about
keeping the village (and its people) clean in the monsoon season. The interview was in a different context entirely, with me arriving in my Western attire to the district office – airconditioned but worn – to share tea and biscuits
with Mr. Pandey and his team of eight.

The officers came across as deeply engaged in developmental issues, and the team showed a genuine concern and affinity for the villages and the villagers they governed. The hillside across the river was a peaceful area, they said, with collaborative people. The villages and villagers were "much the same as they had always been", and they saw no pressing issues except for that of development, of course, which was their area of expertise. Not much was happening regarding development either, they sighed: some connectivity work on roads and mobile phone networks was being developed, but apart from that – nothing.

As the recent devastating flood in Uttarakhand (see Chapter 6) was fresh in our memories, the conversation quickly trailed off to another grave issue in the region and what the officers saw as the direct cause of flash floods: climate change and global warming. The staff expressed sincere concerns about the developments in the nearby industry, the factories that had mushroomed over the last few years and the consequent pollution of air and water. The rising temperatures of the seasons in general concerned them all, a "heating" that, they explained, intensified with deforestation and air pollution. When I asked why the officers had not mentioned waste management and air pollution at the motivational talk, the officer shrugged and said, "It is not so much of it there, so that's no problem". The village, as mentioned in Chapter 2, had been included in an Eco-Sensitive Zone, created as "shock absorbers" between state protected areas like Natural Parks and Sanctuaries (Dash 2009; Srivastava 2011:5) and the pressure of the industrialisation of Indian society at large. In the Shivalik Hills, this particular zone was thought to act as a buffer between the fragile environment of the hills and the heavily industrialised plains regions, which was encompassed in a different kind of zone – a "Special Economic Zone" (SEZ). These zones in the northern territories of Haryana and the bordering southern region of Himachal Pradesh enjoy particularly lax taxation rules so as to enhance economic development and exports (Government of Himachal Pradesh 2013; Department of Commerce 2017; Mohanty and Chandran 2017). This attracts investors from all over the country, and in the first decade of the new millennia, the area saw a massive expansion of industries producing automobile parts, sanitary ware, scientific instruments, pharmaceuticals, electronics, cement, etc. As the Eco-Sensitive Zone prohibits any highly polluting industry, wood-based industries or mining activities (Haryana Forest Department 2016), development in Rani Mājri was ensured to be "ecologically sound" from 2009 (Haryana Forest Department 2016).

Junction 2: governing forest

The forest, as we saw in the introduction to this chapter, was protected by government legislation. One morning, the Forest Guard of the Haryana Forest Department appeared in the village. He had arrived for a meeting with Prakash about the state of the village forest, and, as he was eager to meet the international guests, we were invited to sit down with him for a cup of tea. He told us he made frequent trips to the villages as he handled forest management in four s within the Eco-Sensitive Zone. By rotating whom he spoke with in the villages by "drinking tea at a new house every time", he got a good overview of the situation. He was in a good mood that day and told me that no problems had ever occurred with the people of Rani Mājri. Not every village was so peaceful, he said. He told me that there were often issues with villagers over Khair, the tree particularly cherished for its high-quality firewood and fodder. Khair, as Avani indicated in the chapter introduction, had a high commercial value, but its logging has been prohibited in the region for a century, since enactment of the Punjab Land Preservation Act of 1900, created to save the Shivalik hills from erosion (Haryana Forest Department 2017). I was told that dispensation could be given by the Department of Forestry for selling Khair grown on one's own property, and, around the early 2000s, a few of the major landowners in Rani Mājri had earned a fair amount selling off the Khair trees growing in their fields. As the Khair wood was reddish, it was quite visible in the piles of firewood stored away for winter, so the guard could easily spot any illegal logging.

In Rani Mājri, the forest within the Eco-Sensitive Zone was, I believe, a kind of reserved or protected forest, whilst the forest around the settlement was classified as village forest, which means that all parts of the forest within the village boundaries could be used by its residents. According to Knudsen (2011) the Indian government has operated with three categories of protection since the Forest Act of 1878, and since under the Punjab Land Preservation Act (Haryana Forest Department 2017), the local Forest Guard could correct and punish the villagers with fines or even imprisonment for misuse and illegal logging of the reserved forest. Everyone in the village saw the job performed by the Forest Guard as important – the forest should be conserved, they said, to benefit the next generation.

The young generation was, at least on paper, taught awareness of the environment through public education. On my first trip to the area in December 2012, joining a small group of scientists from the I.I.S.W.C. collecting soil-samples in this village, I had noticed a sign – jolly and fresh – outside the middle school. It read, "Sunflower Eco-Club". I was excited to learn what went on in the eco-club and immediately began to imagine how I could follow its curriculum and activities if I were to choose Rani Mājri and if Rani Mājri would have me.

I found that the Eco-Clubs of Haryana government schools were a part of a 2006 Haryana Department of Environment scheme, where the teachers could "educate the students about a pollution-free environment". This would happen through initiatives like "be[ing] taken to polluting industries from time to time so as to apprise them about industrial pollution". In addition, "Regional officers of the Board have been advised to visit schools and colleges to create awareness among the students about a clean and healthy environment" (OneIndia, 2006).

One of the first things I did after having settled in the village was contact the principal at the middle school of the village, a senior male from a larger city on the plains. He confirmed that the Eco-Club was "for the children to learn about their environment". The activities they had done, he said, were to take the children to the forest and teach them about plants occasionally.

A few weeks later, I began to enquire after pupils at the school in hope of being introduced to the Eco-Club's activities. However, the schoolchildren from the village told me that there were no activities at the local Eco-Club at the moment. In fact, they did not really know what it was. A few pupils vaguely remembered going into the forest once. The adults had no idea either, except for a few of the younger women with childhoods in other villages who remembered an Eco-Club from their village schools. They had been taken on an excursion to villages in the hills, one woman remembered, to learn "to close their taps to save water".

Two of the village's schoolteachers, both middle-aged woman from two different cities on the plains, readily admitted that they had not done much related to the Eco-Club in the village. "We planted some pods for trees, but there hasn't been much more", one sighed. "You know, in the village, so

Figure 3.2 The Eco-Club Sign outside the local school (name of village is removed).
Source: Author.
Image showing a sign with the text "Sunflower Eco-Club".

much greenery is already there". I asked them whether that meant there was no need for an Eco-Club, but instead, they indicated that rural people would have difficulties fathoming the issue – as they practically lived "in the greenery" to which other pupils were taken to learn about the environment. Even in Chandigarh, the other teacher added, "people will not really understand, because of this greenery here" – she pointed to the shrubby hills, which did look quite green from a distance. There were more Eco-Club activities in the schools of the city, where she had been stationed before. City children must learn how plants and trees grow, she said, but she found it to be of no use teaching village kids about environmental issues like global warming.

To talk to the children on matters of global warming or forest conservation might have seemed like an extra-curriculum activity, for which there might be little time. There were other, more pressing issues to be handled here, the teachers said. The prospects of making education available for girls, especially S.C. girls, concerned the teachers more. So did the poor chances of all village children in general to fulfil the national standards necessary to pass on to higher education after primary school. "The higher you go, the poorer are the people", one of the teachers explained, shaking her head with worry. Neither of the teachers seemed proud of the level of education they could offer the hill children, and both expressed quite clearly that their own children attended private schools in the city. They explicitly blamed the government school system and said that if I only knew how bad (*kharāb*) it was, I would also understand how difficult it was to be a teacher in the hills.

Junction 3: *governing soil and water*

In the dryness of summer, a message had been painted on the school building in bright white with blue font, in both Hindi and English: "Save water, secure the life" and "water is life". Neither the students nor the teachers knew who had done it. "The government, I suppose", the teachers said, noting that they had seen the same writing on the schools further down the hills on their way up to work in Rani Mājri.

Water was indeed a source of concern for all villagers, and the local government ran several campaigns in the area to teach villagers how to save it by turning off the tap. Some households, by choice or necessity, utilised government groundwater, pumped and piped from a source further downhill. As we saw in Chapter 2, the utilisation of tax-levied government water was already limited to a few hours a day, and the one-third of households that relied on it (the landless Rajput and Lohar population as well as the S.C. hamlet) were already meticulously saving every drop of their daily allocation. The landholders, whose livelihoods depended on access to local *kuhl* water for household activities and irrigation, were already managing their water and would be constantly guided in how to maximise its profitability by the I.I.S.W.C.

Figure 3.3 The painted letters to the left read, "Jal hai jīvan hai".
Source: Author.
The image shows painted white text on a brick wall reading "water is life".

Rani Mājri was and is, in the I.I.S.W.C. context, a relatively small project, where five uphill villages were covered by a small watershed flowing towards the seasonal river of Ghaggar. When mapping the Agricultural practices of the area around 2006–2007, the I.I.S.W.C. found that the farmers of the watershed still relied on old, traditional food-crop varieties, which lowered their yield as these varieties were riskier than more modern ones due to their susceptibility to insects, pests and disease. The villagers' traditional sowing techniques were also critiqued in the report as they broadcasted the seeds (i.e., scattered them on the field by hand), which led to an uneven distribution of plants and a low germination rate. The farmers were also found not to be weeding their maize crops, nor did they show any understanding of "proper doses of fertilisers, proportion of various nutrients, time and method of application", which would result in "heavy variations in yield levels" (Yadav et al. 2008:29–30). The farmers had also moved away from traditional organic farming practices of using manure as organic fertiliser, a practice the I.I.S.W.C. team would have liked to revive as the combination of organic fertiliser and chemical fertiliser was seen to improve soil health (Yadav et al. 2008:iii). The report thus suggested that the farmers should be introduced to higher-yielding crop varieties, balanced fertilising techniques, line-sowing methods and weed-control measures through farmer training and crop demonstrations; it also advised the I.I.S.W.C. to "raise awareness" of appropriate farming techniques (Yadav et al. 2008:30–31). By rehabilitating the environment, the programme aimed to make production sustainable and to provide a "better livelihood" by enhancing the cash crop ratio for markets and generating regular employment opportunities for 58 persons in the watershed area (Yadav et al. 2008:iii, 100). The socio-economic benefits growing out of more modernised techniques of farming and solidifying the traditional *kuhl* system were considered to be quite high.

Figure 3.4 Workers solidifying the *kuhl* at its base. The construction was a continuation of the work I.I.S.W.C. had initiated to mitigate runoff water.
Source: Author.
Image showing a water pipe being built along a path in a steep hillside.

The renovation of the existing terrace systems and the construction of check dams to stagger the drainage of water and erosion was only part of the I.I.S.W.C. project. Directly intervening in both the physical and social landscape by solidifying the *kuhl*, they also educated the landowners in using the appropriate techniques and tools to tend their land to maximise profit from their yield and how to conserve and maintain the quality of their soil by keeping or planning certain trees and grasses.

Scientists from I.I.S.W.C. would visit the village regularly to check up on the *kuhl* reconstruction process or to register soil samples and record yield and rainwater amounts, orient the farmers about improved crop varieties, hold workshops on farming methods, record produced yield and collect soil-samples for analysing quality (level of siltation, water runoff, soil nutrients, etc.). The team would arrive in pairs or in a group of three, in white cars or jeeps. The researchers, both male and female, were in striking contrast to the villagers, the female researchers, especially – dressing in shirts or jumpers and pants, hair loose and never veiled, sporting none of the traditional bracelets (bangles) or *sindūr* (the vermillion red stripe on the mid-hairline) that adorned married women of the same age in the village. Sometimes the researchers would buy tomatoes or other surplus crops for home-use as they, as well as many others I encountered, considered uphill food both healthier and better-tasting than the plains' produce. At other times, the researchers would take time to sit for a cup of tea and give the farmers advice on what markets would give the best profits for their produce that season. They brought young researchers too, occasionally, interns or degree students, who would do surveys in the village on soil or water management. The students and the researchers would mainly interact with Prakash, who was I.I.S.W.C.'s contact person in the village. I arrived with the I.I.S.W.C. researchers in December 2012 and observed one other German student visiting Rani Mājri with Dr S. L. Arya.

Development trajectories

These junctions do illustrate the sense of the acuteness of environmental issues in the Shivalik Hills. As outlined above, R. C. Gupta, the I.I.S.W.C. scientists and the forest guard were part of a wide range of schemes, initiatives, projects, zones and campaigns directed at the public, especially the rural and poor, to provide the best possible context for development and environmental conservation. What is considered apt development, however, has changed several times throughout history. The post-Second World War optimism of raising the "underdeveloped" (i.e., non-Western) nations up to the "developed" (i.e., Western) nations' level through the transfer of capital, technology and knowledge through the pessimism of the 1980s debt crisis and the failure of a "one size fits all" approach, "development" is continuously revamped in new guises. A classical anthropological, as well as a political ecological approach to these projects of development has been to understand the degree to which they can be said to "succeed" or "fail". However, as James Ferguson's study of a "failed" cattle rearing project in Lesotho, South Africa brought to our attention, such projects should perhaps not be scrutinised by how they might mask or depoliticise other socio-economic factors, to the degree that they even "whisk[...] political realities out of sight" as an "anti-politics machine" (Ferguson 1994:xv).

The more environmentally oriented approaches to development, on the other hand, could, from the 1960s onward, be read in line with Argyrou (2005) as approaches advocating natural conservation as a counteraction to conspicuous consumption. The movement was long associated with a cultural, largely socialist movement in the decades after the Second World War, seeking to protect "nature" from human harm (Cronon 1996:25; Griessler and Littig 2005). The dominant thought was that the more humans there are, the less resources there are. To ensure that resources were not depleted by humans, the local populations in environmentally "fragile" locations had to be guided through expert management, drawing upon decades of Western Cartesian belief wherein nature exists to be managed for human needs (Griessler and Littig 2005).

Ensuring that all humans have equal access to a set of finite natural resources through expert management has thus been the trope of a form of development that has assumed that humans need a set of finite and defined resources to prosper (Robbins 2020). This led to the current sustainability approach, intended to ensure "bottom-up" Sustainable Development, where economic, social and ecological aspects are addressed (see e.g. Griessler and Littig 2005; McEwan 2019).

Ecologic, economic and social (as well as institutional) sustainability should be addressed by all national and international actors (Griessler and Littig 2005; Brightman and Lewis 2017). This largely uneasy relationship of developing and conserving the environment for human needs was most clearly combined in the 1987 Brundtland Commission on Sustainable Development (United Nations 1987); the 1992 UN Rio declaration; and, finally, as a full-fledged international priority in 2015, formulated in UN's Sustainable Development Goals (United Nations 2020); this is the current paradigm, I would say, of how Sustainable Development should proceed in the Shivalik Hills of North India.

Governmentally aided development in India has followed these international turns, based largely upon the premise that the environment is ecologically fragile, that the people are socio-economically underdeveloped and that their unawareness of climate change make them incapable of adapting to those very changes. Soil, water and forest management of the Shivalik Hills was no exception. To improve water reliability, mitigate soil erosion and increase yield and socio-economic development, soil and water conservation measures were executed in the lower Shivalik Hills with "Integrated Watershed Management Programs". These projects are generally credited with replacing "the traditional, fragmented sectoral approach to water resources and management that has led to poor services and unsustainable resource use" and basing water management "on the understanding that water resources are an integral component of the ecosystem, a natural resource, and a social and economic good" (Global Water Partnership 2010). If left in their state of unawareness, the lifestyle and practices of the local population might exacerbate the deterioration of the environment and, in effect, decrease their chances for sustenance agriculture. This will cause an increase to the already problematic number of people migrating from the rural areas of the hills into the cities of the plains in search of jobs. If made "aware", however, the farmer can make the "right" consumer choices, i.e., choices that will preserve the environment, eradicate poverty and contribute to national G.N.P. at the same time. This will fulfil the vison of a sustainable future.

A history of management

The first signs of urban settlement in this area date back to 5,000 B.C.E., when the Harappans, the ancient Indus Valley Civilisation, inhabited the area that today includes the vicinity of Chandigarh and areas along the Ghaggar-Hakra River and its tributaries (Law 2008; Mohan 2012). The area seems to have been populated by humans since. It is hard to say anything about what the landscape of Rani Mājri might have looked like, had it never been governed by petty and grandiose rulers, kings and emperors. The management tradition of *kuhl* systems, for example, has a long history of being related to state management in this region. According to Baker (2005), the hill kings readily adopted *kuhls* as "technologies of power", developed by the Gupta empire in the 4th century C.E. The Gupta kings gave grants of land to temples and Brahman priests, sponsoring the construction of irrigation structures "in order to sanctify and solidify political authority", to strengthen political alliances "and in some cases to transform potential rivals into allies" (Baker 2005:99, 107). Later, the state sponsorship of *kuhls* provided material benefits to the pre-colonial rulers in addition to symbolic ones because "the tax assessed on irrigated land was significantly higher than that assessed on unirrigated land" (Baker 2005:111).

Upon the arrival of the British colonial officers in the region, the gazetteers from the early 20th century, such as the *Ambala Imperial Gazetteer of India* from 1901, describes the general area where Rani Mājri is located as an area of fertile, alluvial loam, intersected by torrents. The torrents "pour down from the hills at intervals of a few miles; and [are] interspersed with blocks of stiff clay soil, which in years of scanty rainfall are unproductive, so that the tract... is liable to famine" (Ambala Imperial Gazetteer 1909:276–277). The region immediately to the southeast, the Morni Hills, was described as having notable amounts of the valuable hardwoods of Sāl and Khair and rich deposits of limestone, and for being so great for game and sports that rewards were given by colonial officers for the killing of tigers, leopards, bears, hyenas, wolves and snakes. According to the elders of the village, there had been much more forest in the hills behind the village up until as late as the 1930s (which was as far back as anyone could remember, see Chapter 2), but below the village there had "always" been fields, at least until the 1750s. This was when the king of Patiala made or planted a forest, according to the village Lambardar. "The kingdom of Patiala had no forests", he said, "so the Raja created one because he enjoyed hunting". With time, this small one grew into a vast forest that, 260 years later, became the forested island of the government Wildlife Sanctuary.

As Fairhead and Leach (1995) argue in their ethnography on deforestation and reforestation in the savanna of West Africa, forest narratives are contested and powerful, and the British colonial forest management has, in both Africa and India, shaped the practices of forest use as well as the relationship between people and forests. Agrawal (2005), Gooch (1998, 2009) and Knudsen (2011) all trace a dramatic colonial forest history. Forest management during colonial times was modelled upon "scientific forestry", meant to maximise benefits from forests, which led to a serious mismanagement of them in the region (Knudsen 2011:299). Especially during the Second World War, the level of timber extraction from the nearby area was momentous (Agrawal 2005). Nomadic people, such as the Van Gujjars (Gooch 1998, 2009) to the north, were displaced with the advancing of modern forestry, and local Agriculturalists were thought to deplete the forest by lopping trees for fodder and burning forest floors for fresh grass and grazing. Rearing goats was prohibited as they were known to contribute to deforestation through over-grazing as far back as 1902, when the British Colonial Government, with a renewed Forest Act (Yadav et al. 2008), closed some lands for grazing and provided various soil-conservation measures (Dove 1992; Agrawal 2005). Post-independence, the Indian government pursued a general trend of privatisation and state control of common property, resulting in an explosive expansion of cultivated areas that led to a "dramatic decline in village commons, especially forests and grazing land" (Baker 2005:15).

The inspector general of forests A. P. F. Hamilton and Colonel Kenneth Mason expressed their concerns early and called for intervention in the *Himalayan Journal* in 1935. The anthropologist Gerald Berreman called for ecological preservation in the 1960s (see Berreman 1978), and the scientists of the I.I.S.W.C., as well as the newer gazetteers, echoed the same narrative in their work of watershed management in the hills (see Kumar and Dahiya 2005). In fact, the deteriorating environment of the Shivalik Hills has been one of the prime causes of state intervention in the region for almost a century. The aim has been to save the Indo-Gangetic plains from erosion and siltation; ensure the well-being of the population of the hills; and slow down economic migration from the hills to the plains, which can increase population pressure in plains cities further.[5] But how can this be achieved?

Self-governance

By and large, these forms of environmental governance draw upon a specific image of the hills as being, at one point in time, forested and rich in plant and animal life. The villagers have variously been depicted as the cause of this deforestation and, thus, will continue to deplete their forest resources unless developed, educated and aided. The message is further reinforced through volunteers from local Non-Governmental, Environmental organisations traveling to the district, who give talks on environmental and/or gendered issues by gathering rural hill-village women to motivate and educate them. International actors were also present, indirectly through institutions and governments funding these activities and processes, and directly by accompanying certain governmental projects. In Rani Mājri, the spectacle that was part of the Clean India Mission was, according to the campaign's webpages, adopting a "community led" and "people centred" strategy, as was I.I.S.W.C.'s approach.

As the advent of a new millennium was celebrated with just the right amount of vigour and anxiety in 1999, many scientists and policymakers were already concerned with balancing natural resource management and development, and were looking for alternative and more "bottom-up" approaches for local acceptance of large scale projects and schemes. It was around this time that the I.I.S.W.C. developed their Hill Resource Management Strategy (H.R.M.S.) methodology. I.I.S.W.C. projects generally operate by involving the local village council, the Panchayat, and the respective Regional Rural Development offices in the physical execution of the projects, which they also did in Rani Mājri. The I.I.S.W.C. staff would pay regular visits to their contact person, often one of the larger landholders with a certain position in village council meetings: Prakash, in this case. In their technical facilitation, as we see, there is a strong dimension of social involvement. When there were problems with water allotment and local struggles over water rights in a village, a "water association", in which the village community had gained responsibility and maintenance of the water and the sharing process, was founded with the intention of ensuring benefit for all landholders – later named "the Hill Resource Management Society" or H.R.M.S. This was, in fact, the predecessor to the Joint Forest Management (J.F.M.) policy in Haryana (Haryana Forest Department 2017). The association turned out to be critical for the projects' success (Agarwal and Narain 2000:11), especially in multi-caste villages or in areas with nomads, such as the Gujjars, where there were competing interests for the water. Now, the scientists told me, a project was more likely to fail because of corruption in the system. The H.R.M.S. method (later also called Integrated Watershed Resource Management [I.W.R.M.]) has since been documented with praise from both scholars and policy makers (Agarwal and Narain 1999, 2000; Lenton and Walkuski 2009). The sustainability, here seen as the continued practice of the projects, seems to rely on the capacity for local self-management.

Governing through "awareness" is an indirect, subtle coercion of people into being, acting or doing. As institutions of development and environmental management gently "nudge" people by rewarding or gently steering the population towards the wanted direction or action, or authoritatively ban, prohibit or prosecute those who abstain or resist rules and regulations, they are effectively moulding new, *environmental identities* (Agrawal 2005; Robbins 2020). Drawing on works from Indonesia (Li 2007), and closer to my own field in India (Agrawal 2005), it becomes apparent that what in fact happens in these junctions is a "proliferation of power" (ibid.). Through new technologies of governance, the state's policies achieve their full effect by turning people into self-regulated accomplices (Agrawal 2005:14, 217). Tracing the history of forest conservation in Kumaon, North India, Agrawal was able to show how the practice of community-based conservation with the creation of local forest councils allowed for local voices to shape regulation conserving their forested area, whilst simultaneously being "shaped anew by the soft hammer of self-regulation" (ibid.:15). In Chapters 1 and 2, I traced some of inequalities between gender, class and caste in Rani Mājri, indicating that the village was far from a uniform unit to be developed. Consider, for a moment, the I.I.S.W.C. watershed management project. In its final year, the I.I.S.W.C. team regarded the project as a success (Dr Arya, pers. comm., 2012). Their researchers had already, in the initial phase, noted "a strong sense of community participation" (Yadav et al. 2008:7), and the team confirmed that the farmers (i.e., the large landholders) had been very interested in extending the irrigated area and increasing their yields, and had collaborated willingly with the officials to do so. However grudgingly the relationship with the more abstracted government was expressed (things that had to do with the larger body of the *sarkārī*, such as elections, corruption and systems failures), all the personified relations were positively valued. The principal of the school and some of the current teachers were highly appreciated, and the villagers would never talk negatively about them, however much they scorned the government for its poor educational system. The forest guard and his job were also generally respected by all castes. For the landholders, the I.I.S.W.C.'s project and its follow-up were very much valued, and Dr Arya's personal involvement was especially appreciated. According to Dipika, my next-door neighbour and friend, Dr Arya was held in particularly high regard by the women of the village. She had taken up part-time residence in Prakash's family's barn down by the road while mapping social demographic elements in the watershed for the preliminary report in 2007. Dr Arya involved the women in self-help groups; spoke up against the men who drank alcohol; and had been the one who, in my neighbour Dipika's words, "made the government do a lot of improvements in the village". As a symbol of how fondly she was remembered, a home-made, stuffed teddy bear, wrapped in plastic to preserve its bright colours, was conspicuously placed in several Rajput and Lohar houses. The teddy was home-made during Dr Arya's sewing lessons, which were much appreciated by the women, who had not learned to sew at home. This indicates, first and foremost, that personified relations of even the most abstracted systems can become good, affectionate even, with time and care.

However, the capacity to forge those relations depends, again, on who you are. As in Foucault (1995) and Foucault and Rabinow (1984), power is also retained within the subject as a capacity to act upon those structures. In that respect, the landholders were perhaps supporting those structures of governance because it reinforced their own positions of power, which we – in the previous chapters – have seen are related quite explicitly to caste, gender and the position held in the local hierarchy. These issues were also emphasized by field-researchers in the I.I.S.W.C itself (Arya, 2007).

The policy and practice of the Indian system of reservation, intended to even out differences between castes, is an intricate and delicate issue. The discussion deserves more space than I can provide here,[6] especially since caste is a collective category and poverty might be very individual, but very briefly: the system of reservation is based on caste identity. In practice, seats are reserved in government education and jobs through the quota system of affirmative reservation, designed to ensure representation as well as re-duce the marginalisation of traditionally deprived groups of society within the categories of Scheduled Tribes (S.T.s), S.C.s or Other Backward Castes (O.B.C.s). As Rajput families were in the general category of Forward Castes (F.C.s), no members of this community received any form of positive discrimination unless they were below the Haryana Poverty Line, but the Lohars and S.C.s were entitled to the quota. Still, the political power of the village was clearly the domain of Lohar and Rajput males. The landown-ing Lohars might have been their equals in wealth and, thus, an important ally in matters of trade and investments, but leadership was dominated by Rajputs. The elected local Gram (village level) Panchayat leader (*sarpanch*) was a Rajput male, and the Panchayat Samiti (district level) leader was also a Rajput. The Panchayat is supposed to hear both sides of a case in disputes, but it has been found elsewhere that it tends to be undemocratic in its failure to provide an ear to lower caste (or non-dominant family clans) and female participants (Madsen 1991:363). Indications of this were noted in this village as well, where the S.C.s had a particularly peripheral role. They did have a female representative in the Gram Panchayat, but the Lohars and Ra-jputs described her membership as merely superfluous. "She signs her name with an X whenever required", the Lohar shopkeeper told me as we talked around the subject of reserved S.C. seats, followed by a short laugh from the men listening to the conversation. In practice, the political influence of the S.Cs were at best marginal. One could thus argue that the upper-caste landholders of Rani Mājri embraced the changes offered to them in govern-ment strategies for development because it served them. The proliferation of power, of which Agrawal (2005) speaks, might have happened because of the social structures already in the village itself. Gold (1999) notes that, in the village of Ghatiyali, farmers did not organise resistance to developments within agro-technology, irrigation schemes and reforestation but rather accepted the conveniences that outside agencies instituted and funded. She describes how this happens, through the farmers' abandonment of a traditional Agricultural ritual held on the auspicious day of Ākhā Tīj. Gold further argues that this abandonment was related to power and social up-heaval within the village itself. With new technology, the farmers could act irrespective of the Brahmins' ritual calendar, embedding the abandonment of ritual in a multiplicity of other kinds of change, such as people wearing other kinds of clothes and eating different sorts of food, and the devaluation of Brahmanical knowledge (see Gold 1999:263, 270). Both Gold (1999) and Agrawal (2005) noted the tendency of the villagers of Kumaon to largely appropriate the structures of "modernity", be they in the form of new strat-egies of forest governance imposed by the government or new technologies of farming – and so did the villagers of Rani Mājri.

Disconnected development

As the improved availability of water and water security due to the watershed-management project had benefited the large landholders, one could perhaps argue that what was good for the landholders of Rani Mājri was good for the village. The marginal landowners and the landless could arguably benefit both ritually and economically from the landholders pro-viding safety through sacrifice as well as a bountiful harvest (see Chapter 4).

Seen from this perspective, the larger landowners earn a fair share selling surplus and cash-cropping, money which is used on health services, build-ing maintenance, farming equipment, transport and education of children, for example. Being a landowner thus seem to provide a better foundation to "adjust" in the future. As made apparent in the preceding chapters, land-holding gives rights to free-of-charge irrigation water and is a class, lineage and caste issue. That hydrology affects social structure in Asia has been well described before, for example, in the state of Tamil Nadu in South India by Mosse (2003). Mosse writes of South Indian water harvesting systems in which "[t]he connections and interdependencies of hydrology interweave with those of caste-class and kinship, business and politics, and generate distinctive patterns of co-operation and conflict" (Mosse 2003:4). However, the village is not a uniform unit of landholders – it also consists of "landless", or utilisers of rain-fed land. As Mehta (2011) notes from research on irriga-tion tank water-distribution in Gujarat, and Baker (2005) from *kuhl* water-distribution in Himachal Pradesh, water is revered by all but controlled by a few. Smaller landholders and, especially, the landless find it difficult to be anything close to self-sufficient and have to pursue other occupations outside the village to make ends meet. Deprioritised by central governments and policy makers in India (Mehta 2011:381), the rain-fed land of Rani Mājri was not encompassed in the I.I.S.W.C. development plan, and neither were the people who relied on it. Instead, those that could have utilised it left it to run its own course and this, in my opinion, reinforces the difference between those with *kuhl* water rights and those without. This, I argue, is be-cause water, soil and forests are persuasively seen as an economic resource only, irrigating the landholder's fields for an increased yield.

But Rani Mājri was a place of unirrigated land too, and rain-fed land was utilised by marginalised high castes and the S.C.s. To be reliant on this rain-fed field alone was quite detrimental to life here as crops can be grown only in the summer monsoon season. The winter rains are unreliable and insufficient for growing the modern variety of seeds, which demand more water than traditional varieties to yield. Deprioritised by central governments and policy makers in India (Mehta 2011:381), rain-fed land is thus seldom encompassed by large-scale state development plans and is instead left to run its own course. Traditionally community managed but state sponsored, the management of *kuhls* has been related to networks of power from early on (Baker 2005:99, 107), thus also affecting the power structures within the village itself. Administrating tax, responsibility and punishment, the family overseeing the old village administration – in Rani Mājri, this would be the Rajput Lambardar – would also be responsible for managing *kuhl* practices and development. To control the two most valuable economic assets in the village, arable land and irrigating water, would thus also allow one to manage relations with individuals or groups with political or economic power in larger cities. When the I.I.S.W.C. improved water-irrigation structures for the landholders by cementing the *kuhl*, they inadvertently cemented the ownership of "free" water in the hands of those landholders. This rendered women in general, and the S.C.s in particular, even more dependent on the (male) landowners and the government's poverty schemes, leaving their voices unheard in debates around the strategies for development in the village.

Some schemes and programmes had provided the "landless" S.C.s with better housing, and a government income (although the work was of low-status and ritually defiling, such as sweeping the main village streets). Simultaneously, the same policies that increased their assets removed just above the Below Poverty Line status, disqualifying them from the benefits such as the government ration cards on foodstuff, redirecting their meagre income to cover food expenses. The constant need for cash income, redirected the S.Cs towards paid labour, not payment in kind by assisting landholders, decoupling the S.Cs from working the land.

When the I.I.S.W.C.'s watershed management project increased the productivity of landholders, who's households benefited from a fertile soil, with crops irrigated by accessible water and protected against insects and pests, their surplus harvest would rather provide cash income from surplus sales. The money generated from surplus income could buy medical services, richer foods, better education and the means to participate in the contemporary Indian society. The increase in cash-flow allowed members of the household to educate themselves and take on work outside agriculture. The current state allowed the family farm to be run by the remaining men and women has made the privileged households of the village better able to adjust to abrupt changes, such as those inflicted by industrialisation, capitalist market systems or global warming. But as the watershed management project was directed at landholders or *kuhl* irrigated land, it did not affect the S.Cs in the same way, who seemed increasingly decoupled from the Agricultural lifestyle.

The difference the watershed management project made, was visible even on the landscape itself. One did not have to walk far to see the effect of the project. In fact, the small Rajput village of Khot was located a short walking distance to the south of the S.C. hamlet and was built as a small, nucleated settlement on a southern ridge. The landowners of Khot, however, irrigated their *kuhl* with water from a different watershed and thus were not encapsulated within the watershed management project. Although the land under cultivation was relatively large, they had only a small fraction of irrigated farmland, provided by an old earthen *kuhl*. Their *kuhl* drew most of its water from groundwater that would gather in a tank under a small grove, constructed by another government development initiative in the 1980s. For some reason that I am not aware of, their local *kuhl* system had not been improved since then. Their irrigation system was thus as vulnerable to drought or years of scanty rainfall as the one in Rani Mājri had been, and the landholders here had a hard time keeping themselves self-sufficient. No buffaloes could be stall-fed here, but there were goats and a few of the Indian-breed cow. Khot had no bus service or road connection that could ease the journey to factories or markets, and the level of education was low across all age groups, compared to Rani Mājri. The settlement had rather recently acquired government-constructed latrines (provided for those Below Poverty Line) and small, unpainted one-story houses of half brick, half clay and mud.

Curiously, however, the villagers of Khot and the S.C. hamlet regarded themselves as being part of Rani Mājri. They shared the same postal code and the same Panchayat, and they had been organised under the same Lambardar. Last but not the least, the Rajputs here came to celebrate the village protector deity Kheṛa Baba (as did the S.C.s), indicating that they could both have been hamlets under one village. As noted in Chapter 2, the agitated S.C. uncle indicated that the S.C. hamlet had been connected to a larger grid. Khot was still partially connected to the Rani Mājri *kuhl* system through a drinking-water pipe, which followed the unpaved road that connected the two villages (but surpassed the S.C. hamlet completely). Before Khot got connected to the Government well water system, which had happened sometime within the decade, the women here would gather and fill their plastic cans at the end of this pipe. During the recurring breakdowns of the Government water supply, the women of Khot would continue to utilise the drinking water from Rani Mājri so the water did still flow from the main village to Khot. It could be that both the S.C. hamlet and Khot were *tikas*, the small hamlets of settlements mentioned in Baker's study of the *kuhl*s in Kangra Valley. In Baker's case, *tikas* often possessed their own irrigation *kuhl*, like Khot, but were still related to one *mauza* or "revenue village". These made up the old fiscal units that the pre-colonial kings used for revenue collection (Baker 2005:55) (see Chapter 2). In theory, the small

Rajput population of Khot – as well as the hamlet of the S.C.s – should have benefited from the same government initiatives as Rani Mājri central village. However, with the current classification of landholder and landless, the S.Cs were excluded from the irrigation benefits provided by the watershed management project, although their settlement was geographically within the same watershed. This strictly geological divide between watersheds, I believe, also segregated the main village with the satellite village of Khot. Piping and cementing the Rani Mājri *kuhl* trajectory meant that surplus floodwater could not access the older peripheral networks of perennial water channels, visible along the path between the central village and the S.C. hamlet. It appears that the S.C.s, at least, were cut off from what might have been a more peripheral system completely because of the solidifying of the main channel. The unintended effect of the watershed management project, I thus argue, was to reinforce the power of a group of landholders that were already at the better end of the spectrum. The old units of *tikas* and the *mauza* might have been dissolved from government records as new administrable units emerged through state bureaucratisation, but they were not entirely dissolved in practice.

"For Avani, as for the rest of the villagers in Rani Mājri, there were clear demarcations between what land or water was to be used, by whom and when, It also became apparent that there were other elements that regulated use and access than the soft hammer of government regulation that structured what is "yours" and what is "mine to use and for what purpose. As the next chapter will show, the use of, and entitlement to land, is regulated not only by physical or administrative borders of the state. To enable a discussion around those elements however, requires making a shift in how to perceive the environment. A less technical, and more phenomenological approach to Rani Mājri's landscape might allow for a richer milieu, the surroundings (vatavaran) of the village. If we perceive the water of the *kuhl*, the trees in the forest and the soil of the fields as something other than a scarce ecological and economical resource, something other than an external environment; if we attempt to perceive it as a place to *dwell*, Rani Mājri takes on a different hue, where clear-cut lines between the nucleated settlement and surrounding villages disappear. The next chapter will attempt to aid that tilt in perspective.

Notes

1 Jiggri, Butti, Bijūl, Sarali and Kaṇte are trees that I have not found any Romanised way of naming; neither have I found their Hindi or English equivalents. They appear here as I heard them pronounced.
2 This might be the White Silk Floss, also called 'Kunth'.
3 *Tang karnā*: to distress.
4 R. C. Gupta had met the former I.I.S.W.C. scientist, Dr Grewal, through this work in the Haryana government with the Department of Agriculture. Upon Gupta's retirement in 2007, he joined Grewal's environmental organisation 'S.P.A.C.E.' (Society for Promotion and Conservation of Environment), founded back in 1997.

5 E.g., Swaran Jayanti Maha Gram Vikas Yojana is a government run, five-year scheme (2016–2021) dedicated to improving conditions in rural villages with the explicit aim of curbing migration to the cities.
6 I refer the interested reader to Gupta's (2005) account of the complexity of caste and class status in contemporary India and how it can materialise in a "consequent clash of multiple hierarchies" (Gupta 2005:424).

References

Agrawal, A.
2005   Environmentality: Technologies of Government and the Making of Subjects. New Ecologies for the Twenty-First Century. Durham: Duke University Press.
Agarwal, A. and Narain, S.
1999   Making Water Management Everybody's Business: Water Harvesting and Rural Development in India. *Gatekeeper Series (SA87).*
2000   Redressing Ecological Poverty Through Participatory Democracy: Case Studies from India. *PERI WorkingPaper Series* (36): 29.
Ambala Imperial Gazetteer
1909   Panjab University, Chandigarh.
Argyrou, V.
2005   The Logic of Environmentalism: Anthropology, Ecology and Postcoloniality. New York: Berghahn Books.
Arya, S. L. and Samra, J. S.
2007   Social and Gender issues in Watershed Development in Shivalik foothill region in India. Indo-US workshop on Innovative E-technologies for Distance Education and Extension/Outreach for Efficient Water Management. ICRISAT, Andhra Pradesh, India,
Baker, M.
2005   Kuhls of Kangra: Community-Managed Irrigation in the Western Himalaya. Seattle and London: University of Washington Press.
Berreman, G.D.
1978   Ecology, Demography and Domestic Strategies in the Western Himalayas. *Journal of Anthropological Research* 34(3): 326–368.
Brightman, M. and Lewis, J.
2017   Introduction: The Anthropology of Sustainability: Beyond Development and Progress. In *The Anthropology of Sustainability. Palgrave Studies in Anthropology of Sustainability.* Brightman M., Lewis J., eds. Pp. 1–34. New York: Palgrave Macmillan.
Cronon, W.
1996   Introduction: In Search of Nature. In *Uncommon Ground: Rethinking the Human Place in Nature.* William Cronon, ed. Pp. 23–56. USA: W. W. Norton.
Dash, D.K.
2009   7 Forests Including Sultanpur Declared Eco-Sensitive Zones. *Times of India.* http://epaper.timesofindia.com/Default/Layout/Includes/TOINEW/ArtWin.asp?Source=Page&Skin=TOINEW&BaseHref=CAP%2F2009%2F07%2F06&ViewMode=HTML&GZ=T&PageLabel=4&EntityId=Ar00401&AppName=1, accessed November 19, 2015.
Department of Commerce
2017   Special Economic Zones. *Department of Commerce. Indian Government.* http://commerce.gov.in/InnerContent.aspx?Id=216, accessed February 22, 2017.

Dove, M.R.
1992   The Dialectical History of "Jungle" in Pakistan: An Examination of the Relationship between Nature and Culture. *Journal of Anthropological Research* 48(3): 231–253.
Fairhead, J. and Leach, M.
1995   False Forest History, Complicit Social Analysis: Rethinking Some West African Environmental Narratives. *World Development* 23(6): 1023–1035.
Ferguson, James
1994   The Anti-Politics Machine. "Development", Depoliticization, and Bureaucratic Power in Lesotho. Minneapolis, London: University of Minnesota Press.
Foucault, M.
1995   The Archaeology of Knowledge. A.M. Sheridan Smith. Reprint. London; New York: Routledge.
Foucault, M. and Rabinow, P.
1984   The Foucault Reader. New York: Pantheon Books.
Global Water Partnership
2010   What Is IWRM? *Global Water Partnership*. http://www.gwp.org/the-challenge/what-is-iwrm/, accessed December 1, 2016.
Gold, A.
1999   Abandoned Rituals: Knowledge, Time and Rhetorics of Modernity in Rural India. In *Religion, Ritual and Royalty: Rajasthan Studies*. Narendra Singhi, ed. Pp. 262–275. New Delhi: Rawat Publications.
Gooch, P.
1998   At the Tail of the Buffalo: Van Gujjar Pastoralists Between the Forest and the World Arena. Media-Tryck. Lund Monographs in Social Anthropology.
2009   Victims of Conservation or Rights as Forest Dwellers: Van Gujjar Pastoralists Between Contesting Codes of Law. *Conservation & Society* 7(4): 239–248.
Government of Himachal Pradesh
2013   Industrial Policy. *Government of Himachal Pradesh*. http://himachal.nic.in/index.php?lang=1&dpt_id=11, accessed June 2017.
Griessler, E. and Littig, B.
2005   Social Sustainability: A Catchword Between Political Pragmatism and Social Theory. *International Journal of Sustainable Development* 8. doi:10.1504/IJSD.2005.007375.
Gupta, D.
2005   Caste and Politics: Identity over System. *Annual Review of Anthropology* 34: 409–427.
Haryana Forest Department
2017   Protected Area of Haryana. http://www.haryanaforest.gov.in/protect.asp Accessed November 22, 2017.
2017   Punjab Land Presentation Act. http://haryanaforest.gov.in/en-us/Act-Rule/Punjab-Land-Preservation-Act. Accessed November 22, 2017.
2017   Joint Forest Management. http://haryanaforest.gov.in/joint.aspx, accessed July 11, 2017.
Knudsen, A.
2011   Logging the 'Frontier': Narratives of Deforestation in the Northern Borderlands of British India, c. 1850–1940. *Forum for Development Studies* 38(3): 299–319.

Kumar, A. and Dahiya, S.B. eds.
2005   Haryana District Gazetteer, Volume II. Haryana Revenue Department.
Law, Randall W. II.
2008   Inter-Regional Interaction and Urbanism in the Ancient Indus Valley: A Geologic Provenience Study of Harappa's Rock and Mineral Assemblage. Occasional Paper: Linguistics, Archaeology and the Human Past. *Indus Project, Research Institute for Humanity and Nature* 11(2011).
Lenton, R. and Walkuski, C.
2009   A Watershed in Watershed Management: the Sukhomajri Experience. Integrated Water Resources Management in Practice: Better Water Management for Development. Mike Muller, ed. pp. 17–29. UK and USA: Earthscan.
Li, T.M.
2007   Governmentality. *Anthropologica* 49(2): 275–281.
Madsen, S.T.
1991   Clan, Kinship, and Panchayat Justice among the Jats of Western Uttar Pradesh. *Anthropos* 86(4/6): 351–365.
McEwan, C.
2019   Postcolonialism, Decoloniality and Development. 2nd edition. New York: Routledge.
Mehta, L.
2011   Social Construction of Scarcity: The Case of Water in Western India. In *Global Political Ecology.* Richard Peet, Paul Robbins, and Michael Watts, eds. Pp. 145–167. USA, Canada and Great Britain: Routledge, Taylor & Francis.
Ministry of Drinking Water and Sanitation
2016   About NBA. Swachh Bharat Mission – Gramin. *Government of India.* http://tsc.gov.in/TSC/NBA/AboutNBA.aspx, accessed November 22, 2016.
Mohan, V.
2012   Harappan Site under Sector 17 – Times of India. *The Times of India.* http://timesofindia.indiatimes.com/city/chandigarh/Harappan-site-under-Sector-17/articleshow/16986832.cms, accessed January 18, 2017.
Mohanty, S., and Chandran, R.
2017   India's Top Court Queries Land Use in Special Economic Zones. *Thomson Reuters Foundation.* http://news.trust.org/item/20170112140906-e5nx8/, accessed February 22, 2017.
Mosse, D.
2003   The Rule of Water. Statecraft, Ecology and Collective Action in South India. New Delhi: Oxford University Press.
OneIndia
2006   Eco Clubs For all Schools. News: India. *OneIndia.* https://www.oneindia.com/2006/06/21/eco-clubs-for-all-haryana-schools-1150846375.html, accessed May 2016.
Robbins, P.
2020   Political Ecology: A Critical Introduction. 3rd edition. Chichester: Wiley-Blackwell.
Srivastava, P.
2011   Guidelines for Declaration of Eco-Sensitive Zones Around National Parks and Wildlife Sanctuaries. *Government of India, Ministry of Environment and Forests (Wildlife Division).* http://envfor.nic.in/content/esz-notifications, accessed December 2016.

Tsing, A.L.

2005   Friction: An Ethnography of Global Connection. Princeton and Oxford: Princeton University Press.

United Nations

2020   Sustainable Development Knowledge Platform. 'Transforming Our World: The 2030 Agenda for Sustainable Development'. https://sustainabledevelopment. un.org/post2015/transformingourworld, accessed June 2020.

1987   Report of the World Commission on Environment and Development: Our Common Future. World Commission on Environment and Development. Oxford: Oxford University Press.

Yadav, R. P., Singh, P., Arya, S.L., Bhatt, V.K and Sharma, P.

2008   Detailed Project Report: For Implementation under NWDPRA Scheme, *Central Soil & Water Conservation Research and Training Institute, Research Centre, Chandigarh.* Ministry of Agriculture, Govt. Of India, New Delhi.

4 Divine jurisdictions

Sardī

In Sardī, the winter season, cold air and fog prevail in the hills and on the plains, and days pass when the sun never manages to break through the thick layer of mist. The season begins with the lunar month of Paush in mid-December and lasts though the lunar month of Māgh, which ends in mid-February.

There is little water in the kuhl, *in the rivers and in the ground when farmers sow the final winter crops – wheat and onions. There has been no rain for months, and the "second monsoon", the winter rains, is awaited. The winter monsoon is different from the summer monsoon in that the rain arrives with less quantity and frequency, and with short and dramatic intensity. A sunny day can quickly go as dark as nightfall, with strong gusts of wind bringing clouds that carry in thunder and lightning.*

The sudden and intermittent rain makes water flow readily in the kuhl *again, and the daily routine of washing the laundry is an easier task. The rain also makes the grass damp and cold, the bedding damp and cold, the air damp and cold and the people damp and cold. They frequently visit the doctor for fevers and pervasive coughs.*

In January, the seeds and roots are all planted in the ground, and all outdoor work is postponed until later in the morning than usual – even the morning bath. At any other time of the year, people take their morning baths before anything else, but in the cold, those who can postpone it until almost noon do. In households with indoor kitchens, more people gather in the warmth from the hearth (ċūlhā) and the sweet cup of tea (ċhai) than usual for breakfast. Even the men, who usually sit separately from the women in the common room, eat in the kitchens for warmth. By turn, people await the bathwater, slowly heating on the coals from the hearth after breakfast. One by one, steaming buckets are carried into the "bathroom stalls" or nooks for the morning bath.

One day, I observed a small and curious metal "stove" used as an extra heating device in the household in which we resided. It looked like an oversized thermo-bottle, with glowing coal from the hearth inserted in the tube to heat the water, which was poured into an outer chamber. It would come in handy, I learned, for all the overnight guests arriving that day to celebrate the marriage of Prakash's youngest brother, Shaan, to a girl from a village on the plains – Rekha. As is customary in traditional Indian weddings, the groom and bride had not previously met: the relationship had been arranged through the alliance of two Rajput caste families from different lineages. That evening, Shaan was ritually prepared by the family pandit, and the family danced and celebrated into the night. In the morning, Shaan and his closest kin left during a ceremony called the "departure of barāt". *This is a procession in which the groom and his family travel to the bride's home village to bring her back with them in a partly serious, partly playful ritual. Staying home with a breastfeeding baby, I missed the ritual greeting which takes place between the families upon their arrival at the bride's village and the wedding ceremony, where the couple, after a ceremony led by a priest, are bound together by a sacred thread and circle the holy fire seven times. I had seen the ritual in the village before, however, when a girl from Rani Mājri was married to a man from another village, and I could easily imagine how the couple departed Rekha's village: encircled by women from her lineage crying and singing sad songs about losing her. The groom and bride now together leave her village, often in a white car, with their family and friends following them in a procession to Rani Mājri, where the rest of the kin, friends and villagers await them. To the beat of drummers and lute players, they walk into the courtyard of their home-to-be. After the customary dancing by the women of the village (in which our neighbour, the toothless and blind widow Lila, made a miraculous effort, considering her old age), the groom and bride enter the house after a short ceremony at the doorstep itself.*

In that small, crammed common room of Bhagwati and Bhupati – the three brothers' elderly parents – all attention was directed at the young bride. Her young and slender body was covered completely in the beautiful, heavy, red wedding sari, and every guest from Rani Mājri was staring at her, eager to get a glimpse of her face behind the red veil. For a second we saw it, as she looked up, flickering her gaze away from the floor. She was pretty, but her eyes tired and distressed; her heavy kājal had smeared a little.

There is one who seems to take no interest in looking at Rekha, however. It is Prakash's wife, Nirmala. After the ceremony at the door, she found a bowl with bright red dye, which she now carries in her left hand. She dips the palm of her right hand in the dye and presses it to the cold surface of the lightly painted cement wall of the house. Her blood-red handprints are left on every wall in every room. The couple is notified, and they follow her as she proceeds out of the house to mark the walls outside. Some women of the lineage gather behind them, leaving the rest of the family and guests behind.

In the crowd of guests, I am not sure what is happening at first, and the couple have slipped out of sight. Asking where they have gone, I am told the women are taking the couple on a "walk about the village" (gāṁv meṁ ghūmnā). *Apparently, they have gone to the small waterfall at the junction of the two ridges where the* kuhl *is diverted. I catch up with them on their return, where they have made a brief stop at the grove. I join the procession there, which now appears to have grown in size, thanks mostly to women from both Lohar and Rajput lineages in Rani Mājri. Nirmala, with the couple and the women trailing behind, now*

Figure 4.1 The red markings a "housewife" makes with the palm of her hand.
Source: Author.
Image showing the offprint of a hand touching a brick wall with red dye.

*proceeds through the village via the main street. All the while, Nirmala marks
the way with the wet, red mixture smeared on the palm of her hand.*

*Through the village and continuing along a cemented path southward, we
quickly arrive at the stairs of the temple of the village deity, Kheṛa Baba. When
we reach the stairs, some women stay behind on the path, but most of them remove their shoes and walk up to the temple. The bride and groom light incense
in front of the temple and perform the prayer (pūja). No women are allowed to
go up the final stairs above the temple, so the groom, the only man in the crowd,
walks up the hill without his bride to make a small offering to the deity enshrined there. Afterwards, the mood is light, and the women start dancing and
singing for Kheṛa Baba, all forcefully chorusing with the "Jai!" that concludes
every verse. As they return to the house, the women reunite with the hundreds
of guests in the eating and dancing of sangīt, and, for the adult men only, the
drinking of "Indian wine": the local whiskey mixed with water.*

*Generally, however, the days in winter are quiet, and as the wedding season
for the Rajputs ebbs, daily rhythm is restored to a regular pace of work, school,
eating and sleeping. In the evening, darkness falls early, just after 6 pm. It
drapes like a carpet over the village, and because of the lack of steady electricity, the area is often covered in complete darkness. Narrow cones of light move
in the dark from small flashlights. The evening meal is taken together in the
common room; the women carry a metal crate with hot coals from the hearth
for everyone – women, men and children – to huddle around. The food is spicier*

than usual, with a lot of chili and ginger, and it is also fattier than usual, with clarified butter (ghī), *both thought to produce heat and energy. Sleeping next to someone in crucial for warmth.*

....

Of the "tour" of the village in the wedding ritual described above, I have found few similar accounts. Sharma (1974), when investigating the degree of collective inclusion to village-shrines (see below), mentions a similar ritual taking place in a village in Kangra in Himachal Pradesh. There is no mention of the red markings made with the palm in Sharma's work, but the procession of the newlywed couple through the village does take place there as an aspect of the traditional arrival-ceremony (*vadhai*) (Sharma 1974:83). Sharma describes this element of the *vadhai* ceremony as a way of "inaugurating" the bride and introducing her to the most important deities of their lineage.

There were also deities that newlyweds in Rani Mājri did not visit, either because of their gender or caste identities or because some ritual sites were associated with danger. I later discovered how powerful these beings of more or less divine origins were thought to be and the agency they wielded in affecting not only the soil or the water that flows through it but the lifeforms that pass, utilise or ignore those sites in which they are worshiped. It does seem that the relationship between those who inhabit and utilise the land is richer and less conventional than the removed bodies of governance would have them be. Accounting for these elements, what then, governs a village?

Deciduous land management

In the previous chapter, it became apparent that defining the *kuhl* users through Agricultural landholding alone led the I.I.S.W.C. watershed management project to effectively exclude the S.C.s, on the basis of their having no irrigable land, as well as the neighbouring village of Khot, on the basis that they were being irrigated from a separate watershed. There were, however, indications that an old form of village organisation, with revenue *mauza* and *tikas,* could have included the two peripheral settlements in the past. Approaching the village with a lens beyond economic (landholder-landless) and the scientific (geological watershed) systems of classification, however, reveals that the settlements are still part of what constitutes Rani Mājri today.

The ritual described in the ethnographic vignette of this chapter, was my first encounter with the divine jurisdiction of Rani Mājri, but how to approach it? One alternative could be to utilize the concept of a "task-scape", as explained in Ingold (2000), which might enable the acknowledgement of non-human actors too, seeing that all tasks according to Ingold are embedded in sociality. As the relationship between the villagers and their deities reveal a more complex topography of Rani Mājri, a more phenomenological reading of the landscape indicates that the village as a place to be developed and conserved would require a consideration not only of landholding and class, but also of immaterial, or non-human actors.

Settled deities

Amid the houses in the central Rani Mājri village, where Rajput and a few Lohar-caste families reside, Hinduism's large and powerful deities appeared on posters and calendars, and in framed images and household shrines. Household shrines are often found in the "common room" of the houses: the room where guests are welcomed for tea or where the family rests, sleeps and eats. With daily ritualised reverence, certain deities are kept close and protect the family from evil or misfortune. In front of the shrines, worship (*pūja*) was routinely conducted in the mornings and evenings before meals, with a short chant and an experienced swirl with incense; in some houses, a jug of water used in the worship is taken outdoors, poured out of its vessel slowly, reciting a mantra with extended/straight arms. This water is referred to as "water of the sun" (*Sūrya kā pānī*) or as "water for Bhagvān".

The deity whose image appeared in most Rajput household shrines was Shiva, also referred to as the "One Above" (*ūpar vālā*).[1] Shiva is worshipped all over India, and everyone in Rani Mājri – including the S.C.s – expressed their devoted faith in him (although the lower castes would not worship him directly, but indirectly through his association with *Kheṛa Baba*, see below). Shiva is an ambivalent deity, both distanced from and engaged in human life. A master of asceticism and fertility, he has a third eye capable of inward vision in peaceful meditation, as well as outward destructive powers. Shiva is, in the context of households, often depicted with his consort, the Goddess Pārvatī. As a couple, Shiva and Pārvatī make a powerful and auspicious pair. Other popular deities were Santoṣī Mā[2] as well as Lakshmi and an image of a lineage Goddess (*kul devī*). The Rajput family, with whom I resided, for example, had a torn and faded black and white photo of a Sandog Mātā shrine as their lineage stemmed from a small village close to Sandog in the Morni hills, where this devī was thought to dwell. Many images found in the house-shrines in the central village were also of gurus. This could be a lineage guru (*kul kā guru*), whose teachings had been followed in the family for generations, or they could be recent additions. Amongst the Rajputs in particular, Guru Rām Rāī (the eldest son of the seventh guru of the Sikhs) was popular, and Prakash and Nirmala had gone to see his Gurdwara in Dehradun on several occasions. Certain Sant Mat[3] gurus, such as Kirpal and Maharaj Charan Singh, were also revered by many.

The S.C.s would worship other deities, or different manifestations of them, than the Lohar and Rajput castes. Guru Ravidas[4] worship was, for example, unique to the S.C.s. His shrine was located a bit further down the road from the stairs to the Kheṛa Baba temple and the village secondary school. In the shade of two enormous mango trees, a Guru Ravidas, looking a little like the head and torso of a man carved roughly in stone, was painted in white and decorated with orange garlands. From that position, he was facing the people coming up the path from the S.C. hamlet. Another deity I only found mentioned amongst the S.Cs was the folk deity Guggā Pir. In this region, Guggā Pir is on par with Bābā Bālak Nāth (Erndl 1998:182) and, according to Fuller (2004:49, 50), is a powerful protector in many North Indian villages. Other than Chandi Devī, Durgā was also found in another form: that of Asha Purni Mata, who I was told had a shrine in the nearby hilly region of Parwanoo.

Moving out of the households, there were several auspicious sites and shrines encircling the village. In the western, lower part of the village centre, where the primary school lies, and the *kuhl* enters the fields, is a small and humble temple dedicated to Hanuman. Hanuman is a monkey deity devoted to Lord Rama and an important God for many farmers. I never witnessed worship here, but I was told it was tended to. A large Hanuman temple in a nearby town a few kilometres down the road was a more popular site for worship, and many Rajput farmers would approach a Brahmin *pandit* there for advice on auspicious and inauspicious timings for Agricultural business. (As Hanuman temples are not to be trespassed by women, I never witnessed any of the counselling.)

At the northern, highest end of the village, where the *kuhl* enters it, is a tiny grove. In 2013, a small ruin and two enormous mango trees provided comfortable shade for both ritual activity and play, and the *kuhl* ran in the open past this site. Beyond this point there was no more settlements before the northern river bed. No temple or structure is present here, but the grove served as a ritual site, mostly for the caste of Rajputs (and on some occasions, I was told, Lohars – but I never observed this). This place could have been a *tirtha*[5] – a site of sacred power. Shiva was one deity who could be approached here. In spring, many young and unmarried girls from the Rajput and Lohar castes made an offering to Shiva during the auspicious time of Shivdhāni (Shivaratri). Dressed up in elaborated *shalwar qamīz* suits, they would make and immerse individual clay *lingams* and a tortoise into the *kuhl* water. The *lingam is* a column with a rounded top, seen as phallic, with the base upon which it is placed representing the female vulva (Caughran 1999:515). The tortoise most likely represented *kūrma*, the cosmic axis, carrying the world on its back (Desai 2009:318).

The other central deity tended to here was Khwaja Pīr.[6] He is, I was told, a local deity that controls water and is considered an important village protector (*rakṣā*[7]). I have found few mentions of Khwaja Pīr in regional literature, and both of the ones I did find are from Hindu worship in Kangra valley of Himachal Pradesh.[8] In the case of Baker (2005), Khwaja Pīr is mentioned briefly as a local deity who can control, guide and calm the local flooded river. In Ursula Sharma's (1974) study on "public" shrines in Kangra, there is a reference to a "Khwajah", as "a deity identified by some with various Muslim personages and saints but almost universally associated with water" (Sharma 1974:72). According to Sharma, Khwaja Pīr's power might manifest in the spring, but his shrine is seldom located there; rather, it is located in more accessible places within the outskirts of the village, where "fields and jungle meet" (Sharma 1974:81). This is because, she believes, as the patron deity of all water sources, he must be worshiped from time to time and, thus, must be made accessible to the landholding castes (Sharma 1974:86). I was told that Khwaja Pīr's *pūja* was performed every year before the onset of the monsoons or in connection with unusual and untimely water-related incidents (see Chapter 6). As far as I know, these would be initiated by a specific household or lineage. In my village's case, I do think this group could include Lohars but not the S.C.s, as was also noted by Sharma (1974) in the case of Khwaja Pīr in Kangra. The influence of Khwaja Pīr's power reached far beyond Rani Mājri, Orpita told me, and she thought he ranked "below Shiva, but above Kheṛa Baba", the village protecting deity.

Kheṛa Baba translates as "lord of the village", a village *Kshetrapala* (*kshetra* lit.: land, soil). As the protector of the village, his presence was regarded as benevolent for everyone residing in it. When conducting my small and regular surveys, I found that Kheṛa Baba was worshiped by all the households, irrespective of caste and/or landholding status, and his birthday was devotedly attended. He had a temple overlooking the village and the fields from a high elevation at the south-western edge of the main village. The temple itself was a rather simple and modest structure. Its walls were left unpainted, but the roof sparkled with white tiles. There was a small stone *liṅgam* at the entrance, revealing Kheṛa Baba's relation to Shiva. It was thought that Kheṛa Baba originated from Rani Mājri, and he was regarded to be in a relationship with the village similar to the relationship between husband and wife. But his origin, like those of other Gods and Goddesses, was unknown to many of the women with whom I spoke. Very few could recite the histories behind the deities present in the village, and most of the information I have on Kheṛa Baba of Rani Mājri comes from the Brahman pandit who assisted my host family in ritual events and celebrations. From the literature, it seems that village deities often lack consorts and might often have a non-divine origin, which also seems to be the case with Kheṛa Baba. As such, he might be seen as a "downscaled" manifestation of Shiva as village protector, making him accessible for mundane concerns – something that, according to Fuller (2004), is well known in Northern India (Fuller 2004:39, 40).

The most popular Goddess amongst the Rajput women was Śerāṁvālī Mātā (She who Rides a Lion). She had no shrine in the central village; instead, she could be approached at the regionally popular Durgā temple in Dhamra, located safely away from settlements, in a relatively large and open space. Śerāṁvālī Mātā is a fierce form of Goddess Durgā, appearing in her form as a mighty slayer of demons (Kinsley 1986). As mentioned in Chapter 2, Durgā is a complex Goddess. She is sometimes worshipped as the consort of a male God and thus inferior, but as Śerāṁvālī Mātā' her *independent form* is worshipped (Erndl 1998:176). The independent Goddess is forceful and

Figure 4.2 Right: The Kheṛa Baba temple of Rani Mājri. Left: The enshrinement of the field deity Panch Pīr.

Left image showing a small concrete building with a square and unpainted base, a small, latticed window and a pointed, white tiled roof. Right image showing two large trees with a decorated base.

almighty, but as she is uncontrolled, she is also potentially dangerous and destructive (Caughran 1999:515). Chandi Devī was another popular form of the "great mother", but with fierce and dangerous aspects. Her modest shrine was also located away from the village centre, in the uphill forest. Her powers would, I was told, keep ghosts and spirits (*bhūt-pret*) away, thus making the forest a safer place to move about. Serāṁvālī Mātā, the mighty Goddess, was approached by the high castes only, but Chandi Devī united all the women in the village – Lohar, Rajput and S.C.s all said they counted on her for protection.

The Goddess in her uncontrolled forms was not the only deity kept at a safe distance from the everyday lives of families in the village centre. On the village periphery, the more unstable, precarious deities dwelled.

The deities enshrined here, in the thicketed forests and on the fields away from settlement, are fierce gods. According to Mines (2005), fierce gods are "contingencies of the Umwelt (environment) over which actors appear to have no immediate control" (Mines 2005:213). To control them, however, one can enshrine them. The practice of enshrining a deity seems to exist because deities (not only ghosts or demons) are dangerous when allowed to roam. Roaming freely, they are uncontrollable, but through "enshrinement" a deity – no matter his or her size – becomes accessible and manageable. Through this, one is, in fact, inviting the deity to settle, "to abandon the lonely and desolate places which are the characteristic haunts of demons and ghosts" (Sharma 1974:82, 83). As might be remembered from the ethnographic vignette at the outset of this chapter, there was another shrine above the Khera Baba temple. Women were not allowed in the vicinity of this shrine as the energy (*śakti*) of the *devta* residing there was particularly dangerous to them. This deity was one of the "fiercer" gods, Sor devta. The S.C.s also had a field deity (whose name I do not know) guarding their small patches of rain-fed land. In a straight line from the temple, in the fields, was the shrine of another "fierce" deity: Panch Pīr/Vīr[9] (the Saint of Five). The power wielded by field deities such as Panch Pīr was considered very strong by the landowners, and upon passing their shrines one would always pay respects by greeting (*pranām*) them. Protection offered by the Panch Pīr shrine whilst working the fields would, according to the Rajput women I accompanied, prevent one from coming to any harm. Their faith in Panch Pīr's control over land would, for example, prevent them from being bitten by venomous snakes, which were encountered in the fields at regular intervals. His goodwill was also crucial for a successful harvest, and after each major one, there would be made offerings to Panch Pīr. Before a new growing season, male landowners (women could not attend this particular worship[10]) would sacrifice the blood of a male goat, killed on the fields next to his shrine, to him.

Placeless beings

Beyond those settled deities of varying ferocities were those that could not or would not be bound, i.e., those that were not controlled, approached, pleased or negotiated with. These were the ghosts and spirits, the feared shadows of lost souls, roaming about, capable of malevolent possessions and trouble. These malevolent entities could also linger and roam freely around the village outskirts and avoiding them required intimate knowledge of their habits.

Between 12 pm and 3 pm, for example, when the sun is at her highest point, people eat and rest indoors for an hour or so. If one were to work outdoors at that time, they might feel dizzy from working in the sun and heat; sensing this weakness, the spirits might possess them. It was known that they could follow you into your home, inflicting disease and problems upon you and your family. Even household animals can be afflicted by the spirits, like Nirmala and Prakash's buffalo, which stopped lactating for weeks because of a spirit harassing (*tang karnā*) the family with acute milk shortage. I vividly recall Nirmala, normally so gentle and mild, repeatedly and forcefully beating her buffalo with a stick in despair after weeks of the spirits withholding her milk. Utmost care should thus be exerted when passing the cremation grounds (*shmashāna*), of which each caste had their own, always located in a riverbed. The Rajput cremation ground lies to the north, about halfway along the path towards the waterfall, where a small path cuts down to the mostly dehydrated river floor. In high summer, when the threat is at its zenith, extra precautionary measures are taken for those who must pass the site. The handful of children from the small Brahman village a few kilometres uphill from Rani Mājri were especially exposed. To reach their village, they had to walk along the *kuhl* and pass the Rajput cremation ground on their way. Thus, in midsummer, when school finished early, they would wait for an hour in one of the Rani Mājri Rajput houses before going home.

Midnight is also considered an inauspicious time, when especially children and young girls should be protected by the benevolent deities of the household. In the dry heat of the night in the middle of the summer, when frequent power-cuts made sleep impossible, many people move outdoors to sleep, which made them particularly susceptible to attacks. I remember well the first nights on which Nirmala and Prakash slept outside the new annexed building. Nirmala started noticing mysterious "tok tok" sounds in the dark around midnight – like a clock but unlike it too. As her husband was a heavy sleeper, he did not notice the sounds. The first night she moved indoors alone, vexed about the situation. The sounds, she thought, were malevolent spirits or ghosts who were known to roam the outskirts of the village in the heat of the night. After the second and third night of this being repeated, she decided to wake Prakash, who also heard the sounds. The very next day, the household's Brahmin pandit was called. The pandit advised certain mantras to be recited at specific times and suggested invoking the deity Ganesha, the elephant God, by placing an image of him over the door. Come evening, Prakash brought a metal crate of smoking hot coal from the hearth. He walked around the new annex with the smoke rising, reciting the mantra. That night, Nirmala felt perfectly safe sleeping outdoors, and in the morning she was relieved to confirm that the "tok tok" sounds had disappeared.

Sometimes, however, the spirits did possess someone, and getting rid of them was a slow and sometimes futile operation. At midnight during Kheṟa Baba's birthday, the villagers engaged in play and song at his shrine, but the festive event took a dramatic turn for Orpita. She was a middle-aged, childless, gentle daughter-in-law of Prakash's household. Married to the second of their three sons, the couple had tried for almost 20 years to get pregnant, with no luck. Assistance from local doctors and ritual experts had given no aid or explanation. Due to her inability to produce an heir, Orpita had been given a role inferior to those of the other two *bahūs*. Still, despite her inferiority she was thankful that her husband had not taken on another wife.

At Kheṟa Baba's birthday that year, the household received an explanation for her misfortune. As Kheṟa Baba's spirit, or *havā*,[11] possessed Orpita for ritual play, he encountered – or was refused by – another spirit that had taken possession of her body when she was a young girl, just before her wedding. The spirit was malevolent and, once revealed by Kheṟa Baba, spurred a ferocious encounter in her body. The spirit made her lose control, scream and throw punches – at her husband, his brothers, her co-wives, anyone who came near her. The malevolent spirit was strong, his possession rooted and forceful, and in panic she (now in control of the spirit) attempted to flee. For all the village to see and hear, she/he was taken back to the house (from which she/he repeatedly tried to escape during the night), continuously speaking in a loud and clearly male voice, with masculine verb forms and in another dialect. She later told me, when she had regained control of herself, and I was finally allowed to see her, that she knew who the spirit was. His name was Taral Singh, an elderly man who had lusted after her in her natal village. She remembered that Taral Singh had wanted to marry her, but she was promised to her current husband. She had, however, in her young naiveté, accepted a sweet from him. With that fatal act, he had taken possession of her, and from that day he had been eating every unborn baby in her womb. She had not been aware of his presence until that day, when Khera Baba attempted to possess her and revealed Taral Singh's presence for all to see. Orpita was kept indoors, on bedrest, for months after this, not working or participating in household activities. Ritual remedies, mantras and advice from pandits, babas and *tantriks*[12] – the ritual specialists that were shrouded in mystery – was hastily acquired. The household also purchased a male goat, which would be the new host of Taral Singh, before it was sacrificed.

As there are specific times when malevolent beings roam, there are also times when benevolent beings control village surroundings. At five in the morning, for example, when it is still dim, and the housewife gets up to tend to the cattle, sweep the manure in heaps and hunch down to milk, there is no timidity, no fear in her movements. This is the "time of God" (*Bhagvān kā time*) or, as some called it, "Goddess time" (*Devī kā time*), the auspicious time of day that begins at 4 am, when the *havā* of the village protector Khera Baba flows though the village, chasing the malevolent beings of the night away.

Auspicious placemaking

In Rani Mājri people would often speak of good (*aćhā*) and bad (*burā*) days, times or places for doing something. Washing your hair, doing your laundry, fasting, performing worship or sowing the fields have to be done at the right time, at the right place and by the right kind of person. If it happened to be a Tuesday or a Thursday, women would avoid washing their hair (and laundry) out of respect to the deities – it would simply be an inauspicious thing to do. Additionally, actions being "good" or "bad" would depend on your gendered body, your aged body and what caste and lineage you belong to – your individual body essence, so to speak. Adhering to the culturally embedded values of auspiciousness and inauspiciousness mattered. As Fuller (2004) wrote on the relationship between deities, people and their surroundings:

[t]here should always be harmony between the deities and the population and territory that they protect and rule over, as well as compatibility between the people and their land, whose qualities are ingested by eating food grown in village fields and drinking water drawn from village wells.

(Fuller 2004:128)

Places are not inherently good nor bad, but they can be made auspicious or benevolent by ritually negotiating with the space you occupy. This happens both by attending to the deities enshrined there and by doing auspicious acts, thereby invoking benevolent powers to protect or guard a dwelling, such as at the inauguration of the newly constructed annex of the house in which we lived in Rani Mājri. For four consecutive days, the Brahmin pandit prayed in front of the four-cornered polygon (*kuṇḍali*[13]) on the floor made of rice and coloured powder, only leaving the building late at night to sleep, then coming back early the next day. On the final day, a ceremony involving the household and the extended lineage was arranged, and only after its completion could Nirmala and Prakash move into the annex.

Figure 4.3 Right: The *kuṇḍali*, an image of Sant Thakar Singh and his successor Sant Baljit Singh, Hanuman and Shri Guru Rām Rāī.

Image to the right show three images: Sant Thakar Singh and his successor Sant Baljit Singh, Hanuman, and Shri Guru Rām Rāī. In front of the images is a squared pattern drawn in white powder.

Just as a house must be continuously kept auspicious by performing the right form of actions in dialogue with the external milieu, so must a village. The importance of such deities for the village identity is widely mentioned in regional literature (Oberoi 1992:376; Fuller 2004:48).

Mines (2005), in her description of ritual life in South India, showed how small and big gods are perceived to use their powers to infuse the soil of a village, and Kinsley (1986) also talked of deities that "are perceived to be not so much transcendent, heavenly beings as beings whose power is firmly grounded in the earth itself" (Kinsley 1986:27). This implies that if the surroundings are managed by ethereal agencies, the governance of people and nature is far from a linear, defined practice to be sorted in government offices. Instead, the governance of humans and land appear with contextually stressed or unstressed borders, upheld in part by ritual practice. Raheja (1988a,b) notes how the ritual capacity of the landholder pivots from the centre to the periphery, passing various entities situated closer or further to the centre, having various capacities for either digesting or passing on inauspiciousness. In Chapter 1, I mentioned Raheja's fieldwork from the village of Pahansu, where the dominant landholder (or king) gives a potentially poisoned gift of *dān* to the high and ritually purest caste of Brahman, who can handle the evil in the gift simply by "digesting" it or transferring it onward in the ritual system (Raheja 1988a:514). The landowner (or king), by giving *dān* ritually, ensures his own and the village's well-being by transferring the evil to the Brahman. The practice of giving these "poisoned gifts" of *dān* also include the village gods and deities, which, just like the Brahmans, can remove or "make far" misfortune and evil (Raheja 1988a:511).

The deciduous land management of fierce beings on the village periphery can thus be bestowed with a function – to encompass or transfer the inauspicious away from the village as a whole. If we see, as in Raheja (1988a,b) and Fuller (2004), the capacity for the deities to encompass evil, we also see the need for their presence in the village periphery, as if they are important to uphold a certain auspicious ritual balance in a centre-periphery model. An encompassing centre-periphery reading of the landscape thus arguably embrace some of the intimacy I sensed there to be, between the ground, the air, the water, the deities and the people.

Negotiating village territories

In this, a village God like Kheṛa Baba appears to be a central deity who ensures Rani Mājri's benevolence and situates the collective village identity (Fuller 2004:146). Kheṛa Baba's birthday is celebrated in mid-summer, and several weeks in advance the women of the village start preparations: sewing him shirts and remembering past possessions, which were considered a grace, granting sacred vision (*darśan*) to his devotees (Erndl 1998:178). This birthday is regarded as a (very) local holiday, when all children were kept home from school, and "village girls" (*gāṁv kī laṛkiyāṁ*) of all ages also returned, from their husbands' villages from near and far, to partake in the celebrations. The Rajputs from Khot also attended, and, with festive and joyous praise, offerings were given at the temple at midday. All the women – except the S.C.s, who were not allowed up the flight of stairs to the temple itself – would then sit cross-legged on the ground, singing and giving in to the occasional possession or "play". A few men, amongst them the pandit used by the Rajputs, would wait and rest in the shade whilst the "play" went on. Come evening, a meal was shared by all villagers in the primary school yard, where all castes, including the S.C.s, who usually kept to their own communities in everyday social life, joined in. The food would be prepared by a few Rajput men in an enormous pot, and people from Khot and Bapūli, as well as S.C. members, served last and seated away from the other castes, came to share the food that was offered to the deity (*bhog*) after the ritual celebrations. The consequences for not adhering to this partial inclusion of the S.C.s in a village feast could be dire and, in fact, provoke retribution by the gods, as Sharma (1974) describes in her account about Kangra in North India. In the case she documents, the lord of rain (called Thakur) withheld rain during a dry winter because not all S.C.s partook in sharing the *bhog* dedicated to him.

This need for symbiosis and balance between benevolent and malevolent agencies also appears in Mines (2005). According to Mines, a South Indian village's protective Goddess stands for the whole village, not metaphorically, like a rose symbolises love, but rather as a synecdoche (a trope in which the qualities of one part of a whole may be said to characterise and suffuse the whole). Mines further suggest that this "encompassing" role of the village Goddess indicates that what is good for the dominating part – the "village people" – is beneficial for all the others too (Mines 2005:33), as if there is a need for symbiosis or balance. This is supported by Fuller (2004), who notes that the inclusion of S.C.s in village festivals such as this is necessary for the village to function ritually because – as with good and evil, auspicious and inauspicious, deities and demons – the pure and impure are symbiotically linked to each other and personify, respectively, "the order and chaos that are, in the Hindu worldview, ultimately inseparable" (Fuller 2004:33). As such, the S.C.s need "to be included so that they can simultaneously be excluded (however partially)" (Fuller 2004:148). Village Gods, Goddesses and fierce Gods alike can exert power that is seen to infuse the soil of the village (Mines 2005:135). As such, the constant negotiation between humans and deities is quite significant for making a place "good". The discussion above indicates that Rani Mājri becoming a good place by certain place-making practices. Built and tended to in the right manner, Rani Mājri becomes, through this ritual optic, a safe and protected place to dwell.

To dwell

Changing one's perception from observing to immersing changes how the "environment" appears too – not as a resource from which we are apart but as a lifeworld in which we dwell, (Ingold 1995, 2000). To depart from a concept of dwelling indicates a sense of place where one perceives oneself as being immersed in an environment. In anthropology, phenomenological approaches have often drawn on the existential philosophy of Heidegger, who classified perception as the most fundamental way in which humans build knowledge or wisdom (Smith 2016). For Heidegger, *dasein*, the "being-in-the-world" (or "there-being"), is a being that perceives the world through activities and commitments whilst being concerned, or aware, about its being (Kelly 2016). Ingold takes a different approach,[14] insisting that "every person is, before anything else, a being-in-the-world" (Ingold 2000:168). Because of this immersion, "I dwell, you dwell" is the same as "I am, you are" (Ingold 2011:10). For Ingold, people develop perceptions and skills in a certain way because their habitus has made them attuned to the environment in which they dwell. Here, he draws upon Bourdieu (2005, 2010) and his notion of "habitus" and practice[15] as humans get to know the world by engaging with it (Ingold 2000:25).

If dwelling in Rani Mājri denotes that cohabitation of place, and places are constituted by actions, processes or events predicated on an understanding of the intersubjective, embodied subject (Whitmore 2018:21), then deities do in fact matter for who and how an environment could be governed. As we can also see from the above, there seems to be a close connotation with the areas governed by malevolent and benevolent beings, and what sites villagers perceives to be for them to utilise and for what. Daniela Berti (2012, 2015), an anthropologist writing about religious and environmental judiciary cases in Himachal Pradesh, notes that village deities are thought to hold property and land rights. From her fieldwork in the Kullu district, she surmised that village deities are thought to rule over specified territories, including the territories of subordinate gods. "Those who live within a deity's jurisdiction feel themselves bound to the deity not only as a devotee to a God, but also as a tenant to a landlord" (Berti 2012:156). In fact, Berti notes the inclusion of divine right to land in Indian judicial cases, with numerous cases regarding environmental disputes pending at the Indian High Courts, which has led to the introduction of "Green Benches" and, in 2010, the creation of a "National Green Tribunal" (Berti 2015:113).

Most of the cases Berti followed in Himachal Pradesh concern complaints about dams, roads, hydropower projects, etc. When assessing the context of environmental conflicts in the region of Kullu, Berti notes that these often take place over territories where village gods are thought to reside (Berti 2015:113). According to Berti, the Indian legal system allows villagers to appeal directly to the High Court with relative ease. Interestingly, the gods are not directly represented in the cases, but they do speak though mediums (2015:113, 114). One example to note is the legal aftermath of the construction of a Hydroelectric project in the forest near the village of Vashist in the Kullu valley, by the private energy and water supplier company Water Miller. This spurred protests from the people of Vashist, who "said that the work would damage the forest where the goddess Joginī supposedly lived"

(Berti 2015:121) and where her shrine was located. The villagers reportedly planned to "call on "Jagti", the parliament of deities, to save the shrine of goddess Joginī" (ibid.:122), and eventually the people of Vashist and other neighbouring villages threatened to boycott the political elections in order to make the government abandon the project. No such conflicts have been noted in the area of Rani Mājri, but the case is still testament to the fact that, even if post-independence land reforms have "caused a considerable drop in wealth for these landowning deities" (Berti 2012:3), the judicialisation of environmental disputes in the region shows that deities still are perceived to exercise power over land.

And if landscape is truly a *world in which we are immersed* (Ingold 2000:207), then the land and the people these deities are said to protect are an indication that Rani Mājri is not confined to the nucleated settlement in the north but also includes the settlements and hamlets of the village periphery. Following Raheja's (1988a) centre-periphery model, as well as considering the experience of the people living in Khot and Rani Mājri, the village is no longer confined to the landholding, cultivating castes of the central village, but in fact embraced Khot and the S.C. hamlet, the forest, the fields, and all the beings that claimed a part of the environment. In fact, all that constitutes *vatavaran* (the surroundings) appear within the same matrix – interrelated and governed by the relationship between humans *and* between the humans and the deities.

However, dwelling should not be taken to denote a romantic, symbolic bond to flora and fauna. Rather, it is one that denotes a relationship of strife, toil, play and routine. This indicates that our experience of the world changes not only within the individual perception but in how the individual relates to a larger social structure of power and hierarchy. In that case, my approach to dwelling is rather "post-phenomenological" (Whitmore 2018). The ability to utilise land, water and forest, as we have seen, is structured both by human relations, forged by kin, caste, state governance, friendships and business – but also ethereal ones. It seems so far that institutions of governance – be they ethereal, ritual or political – often seamlessly overlap in one dimension or another. This means there is a tendency for the control of the surroundings, being governed as a resource for economic utilisation only, to befall those that recognise, and are able to utilise, the tools with which one governs.

As the ritually purer and economically stronger males from the landholding castes were allowed to approach powerful deities and their pandits – as well as scientists and government officers – to consult on planning anything of importance, they were also better situated to draw upon information from relatives, friends and acquaintances in more powerful positions. The "landless" and S.C.s are less apt, I argue, to partake in the "development" or "progress" required to participate socially, economically and politically. This has, I argue, been reinforced by the fact that the government and the landholder castes have operated with a definition of landlessness based on irrigated land only. That implies that, as landowners grew stronger, S.C.s became more dependent both on the landowners and the government's aid, which left them with a marginal say in local negotiations. This is especially true for S.C. women, who were marginalised because of not only their caste but their gender too. As such, the I.I.S.W.C.C's project structure enables what Baker (2005) sees as the state continuing a long history of involvement in local *kuhl* management systems, reinforcing power in the hands of landholders.

In the case of Climate Change Awareness, this becomes particularly acute. With the forest guards, the zones, the Eco-Club, the I.I.S.W.C.'s HRMS and self-help groups, show clearly that the villagers of Rani Mājri were not entrusted with self-governance, unless subjected to a decentralised regulatory rule. I see this as in an example of the shift Agrawal (2005) describes as a move towards more self-regulating bodies of governance. Thus, it seems timely to ask: is climate change affecting how the state produces a certain "environmental subject" suited for governance? I have noted in the preceding chapters that there was a tendency to perceive the villagers – especially women, the S.Cs and the poor – as being *unaware* and *backward*, somehow "lagging behind" progress and development.

Addressing what it is that makes and Indian an Indian, Ramanujan (1989) argued that in India, there is a notion that all things, even space and time itself, are "substantial" (*dhātu*). On the intimacy between humans, entities and their surroundings, Ramanujan (1989:51) states, "[t]he soil in a village, which produces crops for the people, affects their character...houses have mood and character, change the fortune and mood of the dweller". As we saw above, the deities that are enshrined, or choose to settle at certain places, do something to the place and the people who dwell in it. If so, then the immersion of that place affects not who you are but who you consider yourself to *be*. The next chapter will address what kind of awareness the people of Rani Mājri were seen to lack and how that could arguably affect the perception of the self.

Notes

1 Shiva, as well as Vishnu and Krishna, can be approached as a personalised *Bhagwān,* the personal nature of the Supreme, the aspect of the absolute truth in the universe, the *"universal substratum"* from which all beings emerge (Wadley 1977:113, 114) (Fuller 2004:35).

2 The Goddess Santoṣī Mā(tā) might have originated in 1960s Rajasthan and gained a cult after a box-office success movie *Jai Santoshi Maa* in 1975 (Babb 1981:388).

3 Sant Mat: religious direction teaching unity of all religions (Jones and Ryan 2006), see also Juergensmeyer (1995).

4 Ravidas is an important figurehead among S.C.s, especially in North India (Gupta 2005:421).

5 Tīrtha, litt: "a place where one fords a river", indicating both that rivers themselves can be sites of sacred power, as well as sites where humans can cross over from the profane to the sacred realm of the gods (Kinsley 1986:184).

6 Pīr, an Urdu term with Persian referents, meaning "spiritual guide". Its etymology shows remnants of Turkish and Afghan rulers, perhaps especially during the Mughal Empire in the North India/Pakistan region throughout the 13th and 16th centuries (Baker 2005:115; Zoller 2017).
7 Here, note the difference in writing and meaning between '*rakṣā*', translating literally as 'protector' – as in the festival of Rakṣā Bandhan (or *rākhī*) – and '*rākṣas*' – a being with malevolent intentions – actually 'an evil spirit, demon'.
8 However, a spirit or river-God of wells and streams called Khwaja Khizr, also known as Khwaja Khadir, has been identified with a shrine along the Indus near Bakhar, where he is worshipped by devotees of both persuasions (Coomaraswamy 1989).
9 According to Prof. C. P. Zoller, scholar of Hindi literature and linguistics, languages of the Western Himalayas and North Pakistan, Panch Pīr could be *pāṃc pīr* 'the five Muslim saints' but more likely is Vīr 'hero, guardian deity', which Zoller believes refers to 'the five guardian deities'. For readability, I will use Panch, but it is pronounced '*Pāṃc*'.
10 As women were not allowed, I did not witness this ritual, but was given a short recount from Prakash.
11 Havā; litt. air, wind, gas, here, spirit in "wind-form". That deities can take on a wind form is quite common, as has been noted also by Erndl (1998).
12 A *tantrik* is a ritual specialist consulted for problems too grave for conventional approaches of worship. The rituals carried out by him sometimes – but not always – include "black magic" (Frøystad 2016), which made the villagers whisper when talking about approaching him.
13 The *kuṇḍali* is "an astrological chart that combines space and time by spatialising the flow of cosmic time as it is manifested in the movement of the planets" (Gray 2011:87).
14 Heidegger excluded animals as beings concerned about their being in his concept of dwelling, arguing that they might have an environment, but they were "deprived of a world" (Heidegger 1995:239 in Ingold 2011:11). This argument makes Ingold depart from Heidegger's notion of dwelling, and, as such, Ingold proceeds to define his own use of the word, which I follow up to a point.
15 To Ingold, in humans' practical engagement with their environment, they acquire dispositions, a sort of embodied knowledge, adding up to their habitus (Bourdieu 2005, 2010 [1977]). Ingold (2000) reads Bourdieu as saying that the habitus does not so much express itself in practice, as it subsists in it, in a "practical mastery that we associate with skill" (Ingold 2000:162).

References

Agrawal, A.
2005   Environmentality: Technologies of Government and the Making of Subjects. New Ecologies for the Twenty-First Century. Durham: Duke University Press
Babb, L.A.
1981   Visual Interaction in Hinduism. *Journal of Anthropological Research* 37(4): 387–401. Published by University of New Mexico.
Baker, M.
2005   Kuhl's of Kangra: Community-Managed Irrigation in the Western Himalaya. Seattle and London: University of Washington Press.
Berti, D.

2012   Ritual Faults, Sins, and Legal Offences: A Discussion About Two Patterns of Justice in Contemporary India. In *Sins and Sinners Perspectives from Asian Religions*. Phyllis Runoff and Koichi Shinohara, eds. pp 153–172. Brill.

2015   Gods' Rights vs Hydroelectric Projects. Environmental Conflicts and the Judicialization of Nature in India. In *The Human Person and Nature in Classical and Modern India*. R. Torella and G. Milanetti, eds. Pp. 111–129. Supplemento n°2 alla Rivista Degli Studi Orientali, n.s., vol. LXXXVIII.

Bourdieu, P.

1990   The Logic of Practice. Cambridge and Oxford, UK: Polity Press.

2005   Udkast til en Praksisteori: Indledt af Tre Studier i Kabylsk Etnologi. Danmark: Hans Reitzel.

2010   Outline of a Theory of Practice. 25th Printing [1977]. Cambridge Studies in Social and Cultural Anthropology. Cambridge: Cambridge University Press.

Caughran, N.

1999   Shiva and Parvati: Public and Private Reflections of Stories in North India. *The Journal of American Folklore* 112(446): 514–526.

Coomaraswamy, A.K.

1989   Khwājā Khadir and the Fountain of Life, in the Tradition of Persian and Mughal Art. In *"What Is Civilisation"* and *Other Essays*. Pp. 157–167. Cambridge: Golgosova Press.

Desai, D.

2009   Kūrma Imagery In Indian Art And Culture. *Artibus Asiae* 69(2): 317–333.

Erndl, K.M.

1998   Seranvali: The Mother Who Possesses. In *Devi: Goddesses of India*. John Stratton Hawley and Donna Marie Wulff, eds. Pp. 173–194. New Delhi: Motilal Banarsidass Publishers.

Frøystad, K.

2016   Alter-Politics Reconsidered: From Different Worlds to Osmotic Worlding. In *Critical Anthropological Engagements in Human Alterity and Difference*. B.E. Bertelsen, and S. Bendixen, eds. Pp. 229–252. Palgrave Macmillan.

Fuller, C.J.

2004   The Camphor Flame: Popular Hinduism and Society in India. 2nd edition. Princeton, NJ: Princeton University Press.

Gray, J.

2011   Building a House in Nepal: Auspiciousness as a Practice of Emplacement. *Social Analysis: The International Journal of Social and Cultural Practice* 55(1): 73–93.

Gupta, D.

2005   Caste and Politics: Identity over System. *Annual Review of Anthropology* 34: 409–427.

Ingold, T.

1995   Globes and Spheres: The Topology of Environmentalism. In *Environmentalism: The View from Anthropology*. Kay Milton, ed. Pp. 31–42. London and New York: Routledge.

2000   The Perception of the Environment: Essays on Livelihood, Dwelling and Skill. Reissue. London and New York: Routledge.

2011   Being Alive: Essays on Movement, Knowledge and Description. London and New York: Routledge.

Jones, A. and Ryan, J.D.

2006   Encyclopedia of Hinduism. In Encyclopedia of World Religions. http://www.prem-rawat-bio.org/library/encyclos/jonesandryan.html, accessed June 2016.

Juergensmeyer, M.

1995   Remembering a Hidden Past. In *Radhasoami Reality: The Logic of a Modern Faith*. pp. 15–33. Princeton, NJ: Princeton University Press.

Kelly, M.R.

2016   Phenomenology and Time-Consciousness | Internet Encyclopedia of Philosophy. In *Internet Encyclopedia of Philosophy: A Peer-Reviewed Academic Resource*. James Fieser and Bradley Dowden, eds. http://www.iep.utm.edu/phe-time/, accessed September 6, 2016.

Kinsley, D.R.

1986   Durgā. In Hindu Goddesses: Visions of the Divine Feminine in the Hindu Religious Tradition Book. University of California Press.

Mines, D.P.

2005   Fierce Gods: Inequality, Ritual, and the Politics of Dignity in a South Indian Village. Bloomington and Indianapolis: Indiana University Press.

Oberoi, H.

1992   Popular Saints, Goddesses, and Village Sacred Sites: Rereading Sikh Experience in the Nineteenth Century. *History of Religions* 31(4): 363–384.

Raheja, G.G.

1988a   India: Caste, Kingship, and Dominance Reconsidered. *Annual Review of Anthropology* 17(1): 497–522.

1988b   The Poison in the Gift: Ritual, Prestation, and the Dominant Caste in a North Indian Village. Chicago, IL: University of Chicago Press.

Ramanujan, A.K.

1989   Is There an Indian Way of Thinking? An Informal Essay. *Contributions to Indian Sociology* 23(1): 41–58.

Sharma, U.

1974   Public Shrines and Private Interests: The Symbolism of The Village Temple. *Sociological Bulletin* 23(1): 71–92.

Smith, D.W.

2016   Phenomenology. In: *The Stanford Encyclopedia of Philosophy*. Edward N. Zalta, ed. Metaphysics Research Lab, Stanford University. https://plato.stanford.edu/entries/phenomenology/, accessed June 8, 2016.

Wadley, Susan S.

1977   Women and the Hindu Tradition. *Signs* 3(1): 113–125.

Whitmore, L.

2018   Mountain Water Rock God: Understanding Kedarnath in the Twenty-First Century. Oakland: University of California Press. DOI: https://doi.org/10.1525/luminos.61

Zoller, C.P.

2017   Personal Communication.

5   Climate identities

Basant

The Basant *season of spring stretches through the lunar month of* Phālgun, *corresponding to mid-February to mid-March, and into first half of* Ćaīt *in late March. Just over a month long, spring is the shortest season in the village.*

The mornings in spring are increasingly lukewarm as the sun breaks through the layer of fog, making morning baths a less agonizing task. The jug of cool water is poured carefully from a plastic ten-litre bucket, skilfully distributed to cover a whole-body scrub, lather and rinse. After the bath, a thorough cleaning of the mouth follows. This is normally done outside while slowly wandering about or hunching down for an early morning chat. Some brush their teeth with market-bought toothpaste and toothbrushes; others prefer the "dental stick" – a freshly picked twig from the Neem trees in the vicinity.

Closer to seven, a small bell is rung, indicating that Prakash's mother, Bhagwati, has initiated the morning worship (pūja) in front of the household shrine. The first bread is always given to a cow (or a buffalo in the absence of a cow), for Krishna and auspiciousness. The men, back to separate seating, are served next. Sitting cross-legged on the woven beds of the common room, or right on the cool concrete if it's warm, they watch a local news broadcast flicker in dusty colours on the boxed television set whilst eating. The women will, after men and children are served, sit cross-legged on the earthen kitchen floor as the house-wife prepares the bread with clarified butter, salt and chili, and refills cups of tea as they eat. While others get ready for a day of work, in the house, forest or fields, the housewife can finally eat. The leftovers are given to the "house dog", a semi-tamed canine who is never cuddled but always fed. In return, it will – almost certainly, at least if awake and in the mood – bark when strangers approach and keep the bands of macaque monkeys away from the cereals and food grains drying on the roof.

The women, and the men who have no paid labour to do, tie scarves around their heads to shield themselves from the dust and go about their business. The factory workers have already left on their Bajajs or Hero Hondas (light motorbikes), their lunchboxes hanging around their necks. Children walk off to the local primary school or down to the cul-de-sac, where they are most likely picked up by a bus that usually, but certainly not always, runs all the way to the end of the district road. Their hair – long for the girls, short for the boys – never budges an inch. Their shoes are always black, preferably shining – but more often worn (and several sizes too big because they have been inherited or bought to fit for years to come).

*After the cool winter with its occasional rain, there are flowers everywhere. White from the neem, yellow and orange from the lantana bush. Those women who have land hand-till manure into the soil of their kitchen gardens and plant lady fingers, coriander and fenugreek (*methī*), which is dried and used both as a spice and as medicine. The fenugreek, especially, is thought to prevent the increasingly common occurrence of diabetes (*sugar bīmārī*). Tomatoes are still being harvested from the winter's growing season, but there are few of them left now. Soon, tomato-based dishes will give way to white soups, thickened with chickpea flour, with bites of the local radish (*mūlī*).*

*As the wheat sown in winter slowly matures, one must be careful not to mistake the green, fresh, unripe wheat (*kanak*) for grass fodder. The wild antelope (*nīlgāī*) and the swamp deer (*barasingha*) are not such picky eaters, so the men move out to their fields to protect their crop against night-time grazers. From midnight to dawn, the farmers stay awake and try to scare the bovids away with flashing lights.*

With the advent of Ćaīt*, the days grow warmer, and the wheat turns yellow. The first large harvest of the season is mustard. Earlier on, some of its leaves had been harvested for the regional spiced green vegetable dish served with the yellow maize bread (*sarson ka sāg*), but now the plant has turned white and dry. For weeks, there will be the rhythmical hammering of wooden bats on rooftops, beating the mustard bellows for the tiny, black seed to pop out. Some seeds are kept for spices, and the rest are pressed in a machine somewhere and made into oil, used for both food and hair.*

One day, two female teachers who previously had been stationed in the village as part of their rotational duty stop by the household. I remember being called to greet them. I find the women of the house gathered in the backroom, where Prakash's daughters share a double bed. The teachers are sitting cross-legged on the bed and ask me to sit with them, whilst the women from my household sit on the floor. Curious about my business in the village, they begin asking me about myself and my husband. Answering, I take good care to not mention my husband by name but instead refer to him as "Jon's father". I had been taught by the women of the house never to speak his name as, every time I said it out loud, one year would be taken from his life. When the teachers hear this, they look at each other and immediately burst out laughing, proceeding to mock those ways of speaking and behaving that the village women taught me were polite and good behaviour. They proceed to inquire about my hair, scorning the villagers for making me braid it and not allowing me to wear it loose, as city women do these days. They are aghast that I do not dare whistle indoors (or in the village), having been told not to as it could attract malevolent spirits. All of this is ridiculous, they say.

Later in the day, when the teachers have left, I meet Nirmala outside our room. I want so badly to tell her that, no matter what the teachers said, I do

not see her as inferior to them and that she will always have my respect. In my
unsteady Hindi, I try to explain that she is equal to the teachers as best I can.
She just shakes her head, disapprovingly. "They are not like us. We will never
*become equal. We are just people of the soil (*mittī*)", she says, returning to her*
work.

These days, everyone works late refining the ginger, which has been drying
white under the sun for over a month on the roofs. Whole families sit on the
rooftops during the day now, cutting and drying the ginger in manual, webbed
turning-machines, making the air of Rani Mājri rich with ginger debris. The
dust makes the nose bubble with sneezes, and men and women tie scarves over
their faces as they work. Perfectly dried, the ginger is now ready to be sold to a
trader in Delhi for Rs. 2000 a kg. The landowners pool their lot after weighing
and measuring their contributions. After the first part of spring harvest is over,
and the school exams have been completed, many enjoy the short break before
the harvesting of the wheat. Children will be away for school holidays visiting
relatives, and the lack of play and crying and laughter makes the village feel
empty for a week or two.

Around this time, the dry maize and rice-grass fodder from the rabī *harvest*
will come to an end for many of those who have buffaloes and cows, increas-
ing dependence on wet fodder. The decline in fodder affects the lactation of
many cattle, and people begin to complain about lack of milk. In the evenings,
which are increasingly warm, the sounds of petrol-driven threshing machines
that mix dry and wet fodder fill the air. The old, manual ones are quieter,
only creaking a little in their rusty joints. When the threshing machines have
stopped, a longing peacock's call breaks the silence of the night.
....

Though the behaviour of the teachers, recounted above, was my only
experience of such face-to-face rudeness, villagers were often talked about
and approached as backward. This was based on their being perceived as
living in a world immersed in ghosts and customs without scientific explanations. Certain practices and beliefs could thus reinforce a stereotype of the
rural hill villager as in need of awareness, addressed in the various projects,
schemes, campaigns and stunts described in Chapter 3. But what kind of
an awareness is that? What makes it so different from other ways of being
aware, and why does it matter?

Being climate change aware

For an international traveller in the Delhi and Chandigarh of 2013, climate
change was "everywhere". In the high-street markets selling imported global
brands, such as clothes by Nike, Adidas and Benneton, and mobile phones
and televisions from LG, Samsung and Nokia, climate change and its environmental connotations would show up on posters and calendars. On
the street one could pass signs advertising new "eco-friendly" and "green"
neighbourhoods, and on the hotel doors there would be gentle reminders

to save the environment. Reading a local English-language newspaper or strolling around the urban, air-conditioned malls, the words "climate change", "living green" and "eco-sensitive" were certainly everywhere, to be seen by tourists and locals alike as they dined and shopped. Amongst the lower middle-class Indian citizens in Chandigarh with whom I was acquainted, however, the term "global warming" was used more often than "climate change".

"Seven years back we could not sit outside like this, it would be too cold. Now – it is like this, with the global warming and everything", Smita, our middle-aged Jat Sikh landlord sighed, as we sat on the patio in front of her house in the mild December sun of 2012. Her neighbour too, a landowner with a large farm estate further south in Haryana, was genuinely worried. The worst thing about farming these days, he said, was the government's slack taxation on imported foods from China, but global warming on top of that had made his former occupation almost impossible. "It has become all unpredictable. One year it will rain too much, and fields will be flooded, then the next year there is drought. … It's the global warming", he said, shaking his head.

Asking around in the cities about concerns over the state of the environment or climate change indicated that most people thought environmental problems were an issue for sure, but on a global scale and outside of Chandigarh. A middle school teacher in Rani Mājri, in an interview about her Chandigarh home, said that people in the city would never be truly concerned about environmental decay as the environment, from their perspective, looked "all right". The villagers of Rani Mājri, least of all, she said, as they *lived* in the greenery. "Those people who travel out, they will think differently about things", she said, "such as with your snow in Norway becoming less, that is how *you* notice the global warming is taking place". Student volunteers working explicitly to build environmental awareness would argue the same. There were several associations working on environmental issues in the city, like Deeksha, Sapling, the Environment Society of India (E.S.I.) and the international "Association Internationale des Étudiants en Sciences Économiques et Commerciales" (A.I.E.S.E.C.). They were, according to their webpages at least, variously engaged in recycling, tree planting and general awareness-raising through school workshops and "awareness camps". The then-leader of AIESEC's environmental branch, a young, male economics student from Punjab University, admitted that there had been little enthusiasm amongst people. At university, he explained, little awareness-raising was going on. He himself had become "aware" of climate change and its effects on the environment from watching the Indian Discovery Channel. He thought that the "greenery" of the students' surroundings was in fact blinding them from what was happening. "Living here one does not see the problems", he explained.

In Rani Mājri, few actively used or were acquainted with the words *climate change*, *global warming* or *environment*, with a few notable exceptions.

The lack of usage of the climate change word, partly reflect quite well why the villagers appeared "unaware" of climate change. Their lack of education could also explain their apparent lack of acquaintance with the associated issues of pollution, environmental deterioration and global warming. However, environmental pollution was already being talked about as *pradushan* (*pradūṣaṇ*), which became apparent when talking about the qualities that made life good or bad when living in the "greenery".

Life in the "greenery"

People would often say that, despite being rural, they found the village a "good" place to live. For the youth, the village was good because it had family and friends, and a large city at a reasonable distance for education and shopping. For the adult villagers, not that interested in the latest trends or fashion, other aspects were better here than they were elsewhere, such as the closeness of family bonds and the ability to maintain good relations with their deities. In fact, and quite interestingly, everyone would say that they also appreciated the absence of the waste, pollution and smog of the cities. Rani Mājri was regarded as good, for example, because the surroundings were peaceful, or calm (*śantī*), as it lay at a distance from industry, large motorised roads and crowds. Its slightly elevated position in the terrain was another factor. An elevated place would most likely be cool, with chilly air and water, which were considered positive qualities. The presence of wind would also make the sky clear (*sāf sūraj*) with regards to fog or smoke (*dhuāṁ*). The benevolent combination of peacefulness, elevation, wind and coolness was thus used to describe why Rani Mājri was a good place by the people who lived there or to describe somewhere even better. A visiting uncle of one of the S.C. girls with whom I was acquainted, for example, described his home village as "even better... even more peaceful. We have almost no cars, and our water is always cold [compared to here]. In the winter, we do not even need a fridge to keep our food fresh!" This was not disputed. Many of my village friends also expressed a wish to travel higher up into the mountains to "see the view" (or thought that I should go) to experience the coolness of the water and air.

   Tidy surroundings were also considered a positive attribute of village life. Clean or clear (*sāf*) surroundings that were not overgrown or untended and not littered with waste (*kūṛā-karkaṭ* or *kūṛā* for short) were pretty or nice. *Kūṛā* was seen as refuse or waste from household and farming activities, such as plastic, paper, cardboard, cloth, ceramics, etc. The shrubs and undergrowth of the forest, a neglected field or a marketplace, for example, were not *sāf* surroundings. If I remarked on litter gathering on the hillsides, people agreed that it was *ganda* (filthy) and that they did not particularly like to see it gather around the village. But, as there was no waste disposal system, they had nowhere else to throw waste and sweepings than in the *kuhl* and into the shrubby hillsides (which were untidy, anyway). Luckily, the village's altitude helped in removing waste and sweepings. When the monsoons arrived, the masses of water would carry the waste in the *kuhl* down and away, and shrubs would grow over much of what stuck more permanently. As such, Rani Mājri was thus only *temporarily* polluted by litter – it was not a lasting and permanent condition, as was the case with the cities and markets of the plains.

The plains, and especially the cities there, had more of the qualities that made a place unattractive to live in. In fact, those with relatives in the cities, such as a few S.C. families, would bring young city children and babies home to nurture over the summer so as to relieve them from heat and disease. Just as the qualities of coolness, movement of air and calmness reinforced each other, the heat, stillness of air and noisy crowds tended to reinforce each other too. A Lohar farmer with whom I was acquainted would, for example, ask me if I did not find Chandigarh to be too hot and uncomfortable in April. I would agree, and he would say, "The city has a lot of pollution (*bahut pradūṣaṇ hotā hai*). It is better here, here there is air/wind (*havā*), it is clean (*sāf*)". Hazy air from exhaust, factory pipes and automobiles would not blow away, they explained, since the cities of the plains lacked wind. Stillness of air and water were also thought to bring about disease. Summer in the cities of the plains was associated with flies, bad smells and exposure to illnesses (*bīmārī*). Flies, and particularly mosquitoes, were known to lay their eggs in shady and moist locations, typically quiet ponds and backwaters, and in places kept unclean or untidy, which the city markets epitomised. The insects of the cities were also known to carry malaria and dengue, which I was told the village mosquitoes did not. Nirmala noted that the malaria mosquitoes were only in the city "because it is so dirty (*ganda*) there. Dengue is also there, but not here". "Only the itchy ones are here".

However, this state of overall superiority of environmental well-being was perceived to be under constant stress. This became especially clear when talking about food. The Pahaṛi potatoes, for example, were preferred by both city and hill-folk over the plains-potatoes, which were regarded as too sweet. Not only was the taste regarded as better, but the food of the rural, highland kitchen garden was thought to be better for health too. When Chandigarh's middle-class people (or their servants) went shopping for hill-farmer "fruits and veg" in farmers markets, they would always look for hill-produce. I.I.S.W.C. researchers would buy hill-farm produce under the counter for private consumption, and the city-dwelling middle school principal would regularly buy fresh buffalo milk from Prakash. There was also a clear correlation between increased incidence of diseases, such as cancer and diabetes, and the use of chemical fertilisers and pesticides in food production; this was a source of great concern in Rani Mājri. The villagers did not know the science[1] behind such allegations, but many had heard the news stories, such as one in 2013 about 22 schoolchildren in Bihar who had died after their midday school meal was poisoned with pesticide, either by accident or, as rumour would have it, intentionally (CNN 2013).

In the past, farmers would tell me, they had fertilised their crops only with cattle dung. However, the yield had been scarce. During the 1980s, fertilisers and pesticides (lit.: *khetī kī davāī,* field medicine) were introduced by "the government people" (*sarkāri log*). In the beginning the farmers would use only a little – a tad here and a tad there – and the results were impressive and the yield plentiful. But as the years passed, development halted, and the only thing that helped was to increase the treatment. Over the last 15 years, I was told, they had begun to use "very much *davāī*", and many were thought to have become ill from eating local foods. The village shopkeeper told me that someone had recently gotten ill from drinking tea. This was, he claimed, because of the sugar in it, which had come from a plot in the village they knew had received far too much fertiliser. Although a few people I spoke to, like Prakash and his brothers, would shrug at the talk of people becoming ill from the chemicals, calling it nonsense (*bakwās*), many others perceived the difficult situation to have become more acute in the last 15 years (since 1998) and were genuinely worried. A Rajput father and his teenage son, in discussing the matter of chemical fertilisers with me, said, half-jokingly, half-seriously, that, just like the father – who was smoking the thin cigarettes known as *beedis* – the earth had gotten addicted to the chemicals. Like the *beedi*, they were pleasing, but in the long run, they were poisonous. "In the maize, in the wheat, in the ginger; there is this poison – even in the milk that you give your son to drink – the animals will eat the grass fodder from the fields, so the poison ends up in our milk, too", the farmer said. "What happens if you stop?" I asked. The farmer shook his head. "We cannot stop. For years, the soil will not produce enough, and we will have no food to eat". The consequence would be starvation.

Another concern was the expanding industrialisation of the area. One Lohar farmer I talked to about the village's prospects explicitly worried about the factories encroaching on village land. "Everything will be dirty (*ganda*), and all the smog (*dhuwāṅ*) will come with it", he said, shaking his head, perturbed. They were aware of the regulations of the "Eco-Sensitive Zone" in which the village was located, but there seemed to be little faith in it being a permanent protective measure. These worries were connected to demands made by the Haryana Chamber for Industry and Commerce for "free industrial zones" to boost industrialisation and growth. In such zones, industries would be allowed to buy land directly from farmers.

In the village, most, if not all, of those I interviewed about the issue were also worried about the forest waning. The S.C. population tended to blame the landholders of the central village for depleting the forest (as they did for the loss of water). "They have fields and land (*khet aur zamīn*), so why do they have to use the forest? If they stop using it, it will regrow!" an S.C. woman said, grudgingly, when we talked about measures to sustain the forest. The Rajputs, however, would mainly blame the government. According to the old Lambardar, in the days of his youth, the village fields might have given a marginal yield, but the goats would graze in the forest adjoining the village, where fodder, fruits and berries were readily available. The trees of the forest were used for Ayurvedic medicine, but, with the forest disappearing, the interest of younger generations had waned too. Now, he sighed, no one remembered how to get the medicine that was in the trees and the plants.

[Those days] it would rain a lot, on an 8-8-day cycle. It made everything clean. And all these fruits were to be found in the forest: the mango, guava, jujube, the *koronta, medheer* and *kangoo*[2]. There were many animals too: sparrow, crow, deer, antelope, cheetah, peacocks... there is less of everything now. First came the hunters that killed the animals. The government said nothing. Then the British Raja cut the forests down. Now all that is left is the reserved forest.

(Retired village Lambardar, born around 1940)

I also found that the villagers drew a strong correlation between forest and rainfall, and what the elderly Lambardar argues above is in line with the desiccation theory,[3] in which forests "draw" or help produce rain in clouds, and deforestation causes "desiccation" – a thorough drying up. This was a notion most villagers saw as a fact and one they shared with the local Development Officer Mr Sharma and the scientist R. C. Gupta at S.P.A.C.E., both of whom argued that if there is more forest there will be more rain, and with rain comes clouds that cool the earth.

Despite these real, felt concerns, outside a few dedicated offices "no one really cared", to paraphrase one of the scientists at I.I.S.W.C., about the environment of the Shivalik Hills. This reflects how little the environment seems to concern the Indian middle class (and media) in general (e.g., Ranjan 2019a,b). In the Yale-funded *Study on Climate Change Communication* by Leiserowitz and Thaker (2012) this same tendency of being aware of global warming but not linking it explicitly to environmental issues or human emission of greenhouses gases was pointed out. The study showed that, even if people in India respond to surveys saying they observe changes in the overall weather and climate patterns, they do so "without understanding the broader issue of global climate change", and that the level of awareness of climate change and related issues were especially low amongst the illiterate (where they found that 50% had not heard of climate change), contrasted with those who had high school graduate degrees (where only 13% responded that they had not heard about it).

Confirming the findings in Leiserowitz and Thaker's 2012 study, in the Rani Mājri of 2013, climate change in its scientific guise was, indeed, a relatively unknown concept, although global warming was not.

The apparent lack of awareness of the word or concept of climate change could not be purely ascribed to the villagers' lack of English proficiency. Most government officers were at least relatively local, and many spoke the local dialect or the more centralised Hindi when executing campaigns, schemes and projects. To be aware of climate change seems to be about more than being aware of a word. It seems that awareness draws upon a config-uration of processes – global warming, carbon emissions and environmen-tal decay – and their relationship, where humans are active mediators who tinker with the ecological balance as benevolent or malevolent agents.

In 2013, at least, it seemed that exposure to the particular configuration of human emissions-global warming-environmental decay was conveyed through higher education and international media. Climate change as a con-cept thus belonged to well-connected people in well-connected places. The villagers of Rani Mājri were not that. If they went to Chandigarh, Panchkula or the Rajiv Gandhi Technology Park, it was to seek treatment at a cheap government hospital or stand for hours in line at public service offices, or, once, upon my recommendation to apply for a job (but the girls did not even pass the front gate). They would never visit any hotels that informed guests, in English, about the precarious state of the Shivalik Hills, nor would they visit the upscale markets or air-conditioned shopping malls where "eco" and "green" choices were pushed in English in every café, restroom and store. None of the active student environmentalists I met during my stay expressed a wish to visit the hills and instead concentrated their activism on projects in urban schools or on recruiting city children for "awareness camps". It seemed, rather, like no one would travel into the more peripheral villages of the hills unless they had family still residing there. In fact, when we told people in Chandigarh – whether the shopkeeper, the washer man or college lecturers – that we were going to live in a Shivalik Hills village, we were immediately warned of the "backward", "drunken" and "criminal" people there. The local media would regularly reinforce the image of unruly and drunken hill villagers, mostly males, accused of murder, theft and rape, and this portrayal was rarely informed by actual encounters. When I, at one point, discussed the cloudburst and devastating landslides in Uttarakhand with an acquaintance in Chandigarh and told him that, in Rani Mājri, peo-ple thought that the flood had to do with the anger of Shiva (see Chapter 6), he shook his head in disbelief: "There are no educated people in the villages. This is about science!" (lit.: "*Paṛhe-likhe log nāhīṅ haiṅ gāṅw maiṅ. Yah sci-ence ke bāre meṅ hai!*").

Exposure to the concept of climate change would only come through scientists or public officials going to the village for work or research, or through media and public education. As we saw in Chapter 3, teachers re-frained from educating their students about the environment, seeing as they already lived in the "greenery"; when the public officers, NGO workers and teachers were asked why they did not talk to the villagers about these issues, there was a tendency to blame their being "backward", "underdeveloped" and "undereducated".

Television was, by far, the information network that reached the greatest number of people, but the villagers could only afford to subscribe to the "cheaper" packages, which comprised a few local news channels and en-tertainment: religious shows, soap operas or music videos. The radio, like landline telephones, was viewed as a remnant of the past and was never used during my stay. Few would ever see an English newspaper, and, in fact, by 2013, even local newspapers were not available for purchase in the small village shop. Aadit, a young man from the Lohar caste, was far more exposed to the aforementioned concepts than anyone else I met. One day I stopped by to see him and his wife Gina, who had recently given birth to their firstborn, a baby girl. They had withdrawn to their bedroom to watch TV, and I was invited in to see the girl. Soon he began to flick between channels. His family was the only one I was aware of who could afford the more comprehensive package, and, to be kind, he found an English-language channel for me – the Indian Discovery Channel. It was incidentally displaying dramatised images of melting snow and icecaps in the Himalayas. "In 30 years, there will be nothing left", the TV presenter said, in that highly theatrical voice that characterises these popular science shows. "Look", Aadit said, "that's global warming (*global warming haiṁ*)". A newly educated insurance agent from an urban college, Aadit would collect and record rainfall data for the I.I.S.W.C. and managed their more technical equipment in their absence.

Neither his father, however, nor his wife knew anything about "global warming". Very few (if any) villagers born before 1985 seemed familiar with the concepts of "climate change" or the "environment". It was easier for people who worked in development or conservation, it seemed, to talk to the villagers about things they could more easily relate to – the forest, the water, the soil.

The middle-aged taxi driver whom we used to travel to and from the village would often sigh and say he pitied the farmers' children, crammed into buses and over-filled auto rickshaws on their way to study in the plains' colleges. They are being bypassed literally and metaphorically, he said, as we drove past them in our private car. To be Paharị, then, was not only to be from or of the mountains; it was to be different – a certain difference in lifestyle in stark contrast to the perceived "progressive" lifestyle of the urban population.

These stereotypes of the hill peasant have a history, and they interweave with a relationship that the hill villagers and the forest, or wilderness, share. This follows, as I see it, a longer history of urban-rural relationships in India. The British colonial government associated forested (*"jangli"* or *"junglee"*) landscapes with the absence of civilisation and positioned the people living in these hills as the "other" of Western civilised culture (Dove 1992:243, see also Agrawal 2005; Knudsen 2011). The Indian state continued these colonial practices and aligned forest and hill (rural) people firmly on the side of "nature"[4] and plains (urban) people on the side of "culture" (Dove 1992:239). On being Paharị in the mid-1960s, Berreman (1978), in his fieldwork in rural Garhwal (now a district in Uttarakhand), noted how "the Paharị represent a way of... exhibiting a number of distinctive features of social organisation, religion, economy, and technology sufficient to make them seem quite strange, unorthodox, and *jangli* ("uncivilised") to the people of the plains"

(Berreman 1978:330). Even today, the alignment of rural forest people with wilderness and disorder persists; they are perceived as people who do not "obey the norms and laws of the country" (Dove 1992:239).

Their backward state was obviously constituted by their "traditional" way of life; their outdated customs and practices; or, worse, their poor impressions of "modern" ways of life. Rekha, who married into the village during my stay, explained it quite well when she told me about her difficulties adjusting to life in the hills. Not defining herself as a Pahaṛi (she came from a small farm on the nearby plains), she compared her natal home to her new home in the following manner:

> You know, this is a hilly area. People here are not progressive thinking... thinking things like... women should not have employment (*naukrī*). My village is on the plains, our thinking is more like in the city. There, a woman can ride on a scooter and have a job. Life is easier.
>
> (Rekha, 22)

They were seen as lacking many of the amenities of modern lifestyles, such as washing machines, water heaters, air conditioners, accessible markets and broadband internet connections but also awareness of fashions, trends and changing ideologies, be they those of women's rights, plastic waste or ecologically grown foods. That this backwardness had become deeply rooted in their sense of self was expressed quite clearly by Nirmala's comment , and many in the village would describe themselves as "backward" to me. The complex feeling of being "left behind" is somewhat characteristic for the "postcolonial condition", argues Gupta (2000). But being rural also seemed to entail a certain relationship with the past and with what was popularly referred to as "the greenery" – the forest, the hills. This was also, quite interestingly, becoming a part of being Pahaṛi that was idealised by certain parts of society as they were seen as being more sensitive, gentler and more "in touch with nature".

Deprived of science, bestowed with eco-sensitivity?

When I talked to the highly educated and globally oriented youth engaged in charity, non-profit/NGOs or studies at university, some would express a longing for the "simple" rural ways of living that the hills seemed to epitomise. As Tsing (2005) describes provincial Indonesian "nature lovers" to be part of a class formation, young, urban high-class Indians too would also seek experiences in "the wild" by trekking outside the tourist trails in "eco-lodges" off the "beaten track". This was a wanted development from the State Government, in fact, in 2010, a whole Rs. 64.8 million (approximately US$1 million) was set aside for Eco-Tourism activities in the state: log huts, tented accommodation, nature trails, etc. This was done, according to the 2010 Haryana Chief Minister, Mr Bhupinder Singh Hooda, "to bring people closer to nature" (Haryana Government 2010).

Following the global trend of sharing experiences on social media, Indian "nature lovers" also expressed their longing for experiences with "real" nature though large media-sharing platforms like Facebook and Instagram or in personal blogs, like the impressively comprehensive "Hills of Morni" by A. Dhillon (a.k.a, Mornee Tramp). He writes:

> The laid-back life on the hills is the closest one can get to paradise. ... The absence of traffic and the stink of the urban chaos. The clouds that swim into your homes. The simple hill folks. ... It's no coincidence that all the sages and rishis of ancient India sought the refuge of the hills for quietude and for experiencing divinity. There is something about hills that brings calmness to the heart. The experience of being 'in-sync' with nature and God. ... The road to Morni takes you nowhere else. There is no money to be made. No room for the profiteers. But there is romance. There is peace.
>
> (Mornee Tramp 2016)

This portrayal of the hills and their peoples in tourist advertising online seems to have grown over the years. The travel-inspiration profile "HeaveninHimachal" on the social media platform Instagram, for example, was established in 2018 and by 2020 had over 30,000 followers. The profile provides daily updates from Himachal Pradesh through stunning images, promoting the scenic tourist destinations of the region as well as images depicting "proud Pahari culture", such as women making *makki ki roti* (a traditional round flatbread of corn flour) or local dances in scenic environs. Their posts are framed in a contemporary tone of environmental concern and are always signed, "Hills Are Vulnerable, Say No to Plastic, Snacks And Water Bottles. Save Himalayas Don't Litter on Mountains. Help Your Mother Nature, Help Yourself. PROUD TO BE PAHARI [sic]" (Heaven in Himachal 2020).

For this segment, nature has become transcendent and romantic, and is, as such, characteristic of what Tsing finds to be a "key distinction of cosmopolitan youth" (Tsing 2005:130). As the same social media platforms boasting "Pahari pride" were increasing in popularity amongst the handful of older youth in Rani Mājri with access to smartphones (often a Chinese low-market brand), self-portraits of Instagram-friendly village life also appeared amongst youths in the village. Pictures taken in the fields or kitchen; while cooking; or wearing the everyday, traditional *shalwār-qamīz* suits were certainly not part of the scene they wanted to portray. If a camera appeared in those settings, teenagers especially would turn away or veil themselves. The waterfall of the village, however, was a popular site for taking pictures with one's smartphone, to be shared on social media. Here, girls would smile, laugh and strike playful poses for the photographer. These image sessions were well staged, in more elaborate and new *shalwār-qamīz* suits, with perfect makeup, sometimes even posing with their hair in ponytails with a puffed front (called "fancy" style) and in Western outfits, like tight-fitting pants and T-shirts. The girls were always in touch with trending colours, patterns and styles, even if they seldom could afford to wear them. Serene landscapes, such as waterfalls or open mountain views, were considered beautiful (*sundar*) backgrounds that could be photo-montaged into their wedding photos or on pictures posted to their Facebook and Instagram pages. The village youth's self-representations thus largely reflected the idyllic images taken by visitors to the hills.

As the relationship between "progress" and the "past" is complex and undoubtedly ever-changing, and the self-identity of any Shivalik Hills farmer is hard for a foreign anthropologist to capture, I still believe that my observations carry some relevance in noting, perhaps, a changing reflection of what *being* Pahaṛi means to the Pahaṛi themselves.

Nevertheless, people did seem unacquainted with science as well as the grand theories of how the world worked. For months, the young people and women with whom I spent time, especially, would only answer, "I do not know" if I asked why they would give me caraway for my stomach pains, why they added organic manure to a field every fourth year, why one shouldn't walk the trail towards the waterfall in midday sunshine, or why chemical fertilisers were bad. The most common response was that I should ask their husbands, or even the "educated people" (i.e., not them). It didn't occur to me at the time that they saw their kind of "knowing" as representing all that made them "backwards" in the eyes of teachers, scientists and government officers: the educated people that knew the "right" reason behind these things working. When villagers are unable to explain why they use branches from the Neem tree to keep their gums and teeth from decaying, it was just as "backward" as acting awkwardly respectful when meeting an Anglo-Saxon visitor and leaving his or her handshake "hello" hanging in favour of a shy and traditional "namaste". Only after I had spent a good four months in the village did women trust me enough to tell me the about "superstitious" rituals, fear of ghosts, their stances on local political or social issues (it turned out that women, too, had strong opinions on the local parties' shifting alliances with the Bharatiya Janata Party [B.J.P.] and the Indian National Congress Party), the level of corruption in government education and what imported goods from China did to their local economy. One could thus be easily fooled into thinking that the villagers – women and lower castes especially – in lacking the language or boldness to express their knowledge, did not wield any form of theoretical or abstract knowledge of the world. That is a faulty conclusion to draw. I would argue that, even for many males, the reluctance to share their views or theories had more to do with genuine insecurity about whether their knowledge was going to be validated by an accepted authority, such as a scientist, teacher, government employee or Brahmin priest. They were, sadly enough, afraid that I would laugh at them or scorn them for being uneducated – as they were used to.

This ultimately underscores the role of what kind of knowledge is needed to be aware of climate change. What "kind" of knowledge should be allowed to inform your perception of the world – thus also the *whys* and *hows* of global warming – play a central part in crafting what could be called "climate change identities". This relates to how people's, especially rural or village people's, knowledge about the environment often appears polarised in a dichotomous relationship between "scientific" and "traditional" knowledge. These dichotomous forms of knowledge continuously reappear, both in scientific studies on Indian "environmental awareness" and in everyday Indian conversations, politics and policy. Here, the debate around whether Hindus have an essentially different concept of pollution than "Westerners" because of a conflation of ritual and physical dirt must be briefly addressed.

Older people in the village would use the word *"gande"* (lit.: filthy, dirty, foul), not *"pradushan"*, when describing garbage or smog from factories. Incidentally, they used the same word, gande, when describing the S.C.s, soil or nail clippings. But does the use of "dirt/filth" extend to conflating ritual and environmental pollution? My own research material does not indicate such a thing. On a practical level, in Rani Mājri at least, people seemed quite clear on the difference between ritual and physical pollution. Here, I echo Haberman (2006) who, in fieldwork amongst worshippers of the environmentally polluted Yamuna river, argued that that if these are at all related, it is because the environmental degradation of the Yamuna is both an ecological problem and a religious crisis (Haberman 2006:1).

The dichotomy, however, of "spiritual or religious knowledge" and "modern or scientific knowledge" appears mainly in two guises. The first is arguing that the Western, scientific kind of knowledge somehow eradicates the local, traditional kind by imposing a different sort of rationality, one that is context-independent and more universalist. In some cases, this argument has been an important tool for vocalising marginalised people's right to be heard. Argyrou (2005), for example, pointed out in his critical book on "environmentalism" that the dualism has enabled certain people, on the basis of their lifestyles as "non-Western", to portray themselves as "victims of a monumental historical misunderstanding" and "as living embodiments of an urgently needed ethic of respect for nature, as repositories of a simple, yet profound wisdom that the West has long lost in its heedless march for progress" (Argyrou 2005:viii). In these more extreme cases, certain archetypes of knowledge might even appear more "natural" to certain cultures than others. For example, some studies argue that the Hindu or Indian (i.e., non-Western) mind is "traditional" in its essence. Nelson (1998), for example, indicates quite clearly in his edited reader on Indian religion and its relationship with ecology that the essential Hindu tradition is entirely "eco-sensitive". He further argues that the "traditional" people of India live in a "Hindu universe, [which] though under siege, is still very much alive" (Nelson 1998:7). This rhetoric carries much sentiment for Hindu nationalists and the blurring of the boundaries between religious environmentalism and Hindu nationalism is well known (see Nanda 2004; Mawdsley 2006). Here, I appreciate the works of Scott (1998) and Turnbull (2005), as well as Hastrup

(2013), who argue that abstract and practical forms of knowledge are messier, more closely related and more complex in their archaeology than how they first appear. Although I sympathise with the notion that certain forms of knowing the world are indeed vanishing as generations that held them die out and that new generations are unable to reproduce them because the contexts in which they grow up are so different, I am deeply uncomfortable with the essentialisation of Hindu eco-sensitivity or what we might call the "oriental ecologist approaches" (Bruun and Kalland 1995; Huber and Pedersen 1997). The idea that environmental perceptions are more "spiritual" in the religions of the "East" than in the Protestant and capitalist nations of the "West" is as naïve as it is reductionist. Even if the Hindu deities do dwell in the Shivalik Hills, there is no guarantee they would want to preserve the environment for human benefit, nor do they seem to align with the current discourse of Sustainable Development (see Adams 2009). Berti (2015) describes a case in which a North Indian deity's dwelling place was declared part of the Great Himalayan National Park in 1999. The deity, through a human medium, disagreed on the park rules forbidding all access to the forest in order to preserve it (Berti 2015:120). The conflict of interest of the human medium aside (which would in any case only underscore the will of the local Hindu population), Berti's main argument is still that there is no guarantee of a god-controlled nature concurring with ecologist arguments. In some cases, a god-controlled nature might very well be "the main argument that prompts villagers, with their gods, to oppose the way an environmental protection policy is implemented" (Berti 2015:13).

The second guise of a polarised view on knowledge could be used to argue that people's practices and perceptions about forests, weather or farming are "traditional" in a destructive sense, where their "unawareness" leads to a detrimental effect upon the environment.

Only after being "modernised" and "aware" following institutional transformation can they be trusted to self-govern (Adams 2009). This relates quite explicitly to the kind of awareness propagated by the government in the "junctions" described in Chapter 3. At its worst, the segregation of knowledge into "traditional" and "modern" can serve to simplify and essentialise "Westerners" and "Easterners" alike. The consequence might very well be that people appear as alienated species with competing and conflicting ways of knowing the world. It is time to look at climate change as a concept with a certain configuration in history in a more discourse-oriented approach.

Climate change as a discourse

As I outlined in the introduction, knowledge about climate change, what it does and how to best address it is transmitted through a myriad of sources. It passes through State Government offices and through United Nations roundtable discussions, editorial offices and NGOs, to mention a few. In these institutions, climate change data on ocean acidification and increased global temperatures depart from being numbers in charts. Meteograms, graphs and models here entwine with corporate and diplomatic interests, state self-determination rights and human rights. In a bewildering nexus of interrelated issues, climate change is thus increasingly addressed as a wicked problem, not only because there are so many unknown factors to the ecological dialectic of how species depend on external factors but in that it ceases to be a scientifically objective measure of the meteorological condition of the globe and becomes, rather, a social, political and economic issue. One expression of this is how climate change appears more and more as a meta-discourse, encapsulating and surpassing the complex relationship between development and conservation, and providing, instead, an encompassing worldview of how the world "best" proceeds. It is here that scientific facts are caught up in *climate change discourses,* embedded and reinforced by decades of conflicting interests and visions of how societies should or should not develop in relation to others. An example of how that happens appears in Myanna Lahsen's work on how this discourse of climate change gets tangled up in the more complex global relations of trade, national sovereignty and policymaking in Brazil (Lahsen 2009). Through tracing forest resource management policy implications in the context of climate change, Lahsen indicates that knowledge of climate change as a physical, real process with the potency to see into the uncertain future has become more than just a scientific fact – it is also becoming an essential element of political rhetoric and for moving substantial amounts of money through the funding and financing of institutes, projects and business enterprises. Climate change as discourse wields structural power.

Addressing climate change as a discourse is important, as Leichenko and O'Brien (2019) argue; addressing the discourse means addressing the understanding of climate change as a problem and, subsequently the appropriate action to be taken in resolving it. In his *Archaeology of Knowledge*, Foucault (1995) explained his use of discourse in referring to knowledge being caught up in a system of references, a group of statements that belong to a single system of formation (ibid.:108). As Foucault's notion of power is elusive and "not located in a privileged place or position, but in the relationship itself" (Deleuze 1999:26, 27), discourse, too, is to be found neither here nor there but in the relations and actions that derive from it. The discourse then manifests "as practices that systematically form the objects of which they speak" (Foucault 1995:49). Echoing Foucault, many discourses can thus be said to exist at the same time, competing in a discursive field. To argue that there is a discourse of climate change, then, still allows for people to disagree on the technicalities around its causes and consequences as long as they participate in the same "conceptual field" (ibid.:126). It does not have to be the case that certain statements are hushed into silence on purpose; rather, these exist under "the positive conditions of a complex group of relations" (Foucault 1995:45). These relations can be seen, as Foucault sees them, as primary, i.e., independent of discourse (in the case of climate change, that the globe is warming at unprecedented rate), and secondary, i.e., what can be said about it (for example, why climate change is happening and what ought to be done about it). These secondary relations depend on discourse, a certain contextual web of "true" statements. The discourse of climate change, then, will consist of a particular and presupposed way of looking at the process of global warming – one that defines what is possible to talk about as the primary driver or ultimate solution of climate change. How the local discourse of climate change unfolds, in other words, shapes the awareness of the problem.

Let us pause here. Climate change as a phenomenon is defined by experts and takes place within a certain discourse where humans play a central role in mitigating or increasing its effects upon the globe. Climate Change Awareness is something that is being campaigned for by a wide range of actors. The UN has been one of the foremost institutions asserting the need for climate change knowledge and policy with all governments, and they have been particularly influential in shaping the discourse of climate change through their Climate Change Panel Environmental Program and their own non-profit communications bureau, GRID, which is dedicated to communicating environmental and Climate Change Awareness in political and decision-making networks (see, for example, GRID-Arendal 2020). The U.N.E.P. also provides lessons in the field of global awareness-raising, with a programme that received the 2007 Nobel Price Prize (with Al Gore) "for their efforts to build up and disseminate greater knowledge about man-made climate change, and to lay the foundations for the measures that are needed to counteract such change" (The Nobel Foundation 2014). Although a leading actor, the UN is not alone in generating international discourse on environmental issues. Individual states, too, wield much of the same power, accepting or declining advice from the UN as they formulate their own climate policies and build political careers on environmental policy or climate change adaptions. What is all this awareness for, one might ask?

A few years ago, there seemed to be an assumption among scientists that the reason why people did not change their behaviour to more sustainable living was their lack of knowledge. The knowledge argument was found insufficient by Norgaard (2011) as explanation (she suggests instead states of denial), but the global community still seems adamant that knowledge – or awareness, rather – is key. The European Climate Adaptation Platform (2015) states that the intention of Climate Change Awareness campaigns is "to achieve long-term lasting behavioural changes". All over the world powerful actors within states, religions and markets are directing their efforts at changing human behaviour. But what kind of behaviour do they want? Is it a different kind of behaviour than the kind that, in Rani Mājri, appeared as explicit instructions on how to live their lives, ranging from how best to do their farming to how to save water when brushing their teeth?

In the discourse of climate change, as in both the development and the environmentalist discourse, people are expected to gain what anthropologist

Kay Milton (2002) calls a "planetary concern" (Milton 2002:170), aimed to mobilise, on behalf of overall humanity, action that produces a sustainable future for humans on planet earth. This is no less than what Tsing (2005) called a "universal dream", the "something we cannot not want" – another example of which is world peace (Tsing 2005:8). In this case, the aim is to solve problems implicated by climate change, mass extinction or pollution. The idea of the universal, however, is thought to belong to a Western mind-set and is "contrasted with more local or indigenous forms of knowledge and being" (Tsing 2005:8) inherent in a modernist paradigm.

As we have seen above, in India, the discourse on climate change does not appear without historical context; neither does it appear independent from other discourses. It is, rather, seen to draw heavily on discourses of development and environmentalism described in Chapter 3, enveloped within the one of climate change and enforced through the trope of Sustainable Development (Griessler and Littig 2005). As Dubash (2012) argues in his edited volume on climate change and policy, both the developmental and environmental discourses can be said to have been appropriated on the state level, engulfed by climate change as one encompassing approach (ibid.:7–9, 16). As Argyrou (2005) argues, educating people on adopting an "environmentally aware" state is very much related to notions of modernity. This relation, he suggests, produces "the same sort of global power relations and the same sort of logic that marks the modernist paradigm at its core" (Argyrou 2005:x). Inherent in the Western, scientific discourse around climate change, with its associated implications and mitigations, is a notion of how humans should and ought to relate to the environment and our fellow beings. These discourses, as we have seen, firmly place humans "at twos" with nature. This is rather ironic as, in line with Ingold (1995, 2000, 2011), Chakrabarty (2009a,b), Hulme (2009) and Rosa (2019), for example, disengagement (or lack of change in behaviour) might rather be looked at as being a problem of humans perceiving themselves as being "removed" from "the climate" through insisting that humans are disconnected, as it were, from their environs. This detachment leaves dispassion; thus, no true agitation can make humans change their current unsustainable behaviour. This, argues Ingold, is central to the contemporary environmental crisis because it entails a humanity that no longer immerses itself within the world but, rather, removes itself from it. This act of removal, however, must be taught.

As discourse affects the capacity of a system's structuring power on the individual, can climate change as a discourse change how people think about themselves? As we have seen, the sole reason for awareness-raising in Rani Mājri drew its force from the villagers' lack of scientific, technical and rational forms of knowledge. According to Agrawal (2005), a transformation happened through the Kumaonis self-regulation, one that changed how people "think about their actions, positively or negatively, in relation to the environment" (ibid.:17), thus crafting an "environmentality". To Agrawal, "environmentality" was what let the colonial and post-colonial Indian state shape or form "environmental subjects": "people who have come to think and act in new ways in relation to the environment" or "for whom the environment constitutes a critical domain of thought and action" (ibid.:16). To understand how this is so, Foucault's notion of "governmentality", upon which Agrawal's concept builds, is useful here. Foucault (1995) was intrigued by how power works upon individuals and groups as subtle coercion, making them do what is wanted of them, often without knowing that they do, as a "microphysics of power" (Foucault in Foucault and Rabinow 1984:183). "Governmentality" refers to "conducting oneself" according to a kind of governmental rationality, exercised by the state to improve its populations. In line with Foucault's notion of power and the body, Agrawal then proceeds to argue that this particular kind of "environmentality" might even change how people "think about their actions, positively or negatively, in relation to the environment" (Agrawal 2005:17). It could also very well indicate that campaigns on "awareness" have the potential to craft climate change identities.

I argue one would be wiser to see knowledge of the world in which we are embedded as a "motley" or a "bricolage" and not as opposing, polarised worldviews. This is also Scott's central argument (Scott 1998). As in Scott, the concept of *mētis* (practical) and *techne* (theoretical, abstract) are not in fact competing versions of knowledge but rather knowledge-practices that sometimes act regardless of techne and sometimes in accordance with it. Thus, knowledge can be both "theoretical/abstract" and "practical/concrete" at the same time. These two forms of knowledge – the abstract and the practical kind – are always related and always interacting, in all societies, at all times (Scott 1998; Ingold 2000, 2011; Turnbull 2005), but to various degrees at different times and in different contexts. This dissolves the concept of there being an indigenous, traditional or modern knowledge. As Cruikshank (2001) observes when looking into climate change knowledge amongst the scientists and locals of Tinglit, North America, if "local" knowledge appears vague, subjective, context-dependent and open to interpretation to a scientist, then the local Tinglit "are quite likely to characterise science in similar terms: as illusory, vague, subjective, and context-dependent, and open to multiple interpretations" (Cruikshank 2001:390). Cruikshank's approach indicates that all knowledge essentially is "local", as argued by Turnbull (2005). In line with this, knowledge is seen as "motley" or as an "assemblage" of many components. We would be wiser to see knowledge manifest, or take shape, "in the moment of practice" (Scott 1998:332). This way, it becomes a sort of "bricolage", recombining all the time and continuously building on itself, like a coral reef. It might branch out here and there, sometimes adjusting and adapting to context, but in practice, it is knowledge all the same. Like Turnbull, then, we should see Western science as an ideological marker in the creation of the "other" (Turnbull 2005:7), where scientific knowledge underpins the celebration of "modernism" as supposedly synonymous with development and social improvement (ibid.:7). If we now return to the premise of Climate Change Awareness campaigns, my ethnography shows that the villagers of Rani Mājri knew very well and wielded their knowledge of the dialectic process between human action and inaction, in addition to the accompanying changes in weather, seasonality and environmental decay taking place. Although, arguably, the scientific idea of climate change has diffused rather poorly in Rani Mājri, it was certainly not because of a lack of the capacity for "abstract thinking". It, we have seen, had more to do with how their "traditional" way of life was approached across a dissonance between humans and their surroundings that the people of Rani Mājri themselves did not experience. If the villagers saw themselves as "backward", it was not because they did not wield theoretical or abstract knowledge. This seems to rely more on their knowledge of the world not being validated by an authority figure (i.e., a scientist or any other "properly educated" person, such as a Brahmin *pandit* or a college teacher).

What they were unaware of was a specific discourse of climate change – the biophysical or technological explanation of global warming and the increased politicisation of this awareness. They were also unaware of the peculiar segregation made in this technoscientific discourse between culture and nature, humans and their environment as, in Rani Mājri, there were strong indications of their correlation. This also gives room for an alternative take on what climate change is, why it happens and what to do about it. Would we comprehend their perception of climate change better if we approached it as having something to do with socio-political inflictions on the ecology and environment, *and* with social relations to those actors of good and bad nature around us? The next chapter will attempt this.

Notes

1 There is a ready availability of medical studies pointing out the effects of pesticides and fertilisers in farming in India, which have contributed to this fear.
2 I have not found any translation for *koronta, medheer* or *kangoo*.
3 This is a matter that has been in dispute for many years, and evidence for this theory is contradictory, but there are studies that point to its validity: see, for example, Makarieva et al. (2013).
4 On nature as a product of Western tradition, see Cronon (1996a,b) and Descola (2013).

References

Adams, W.M.
2009   Green Development. Environment and Sustainability in a Developing World. 3rd edition. Oxon and New York: Routledge.
Agrawal, A.
2005   Environmentality: Technologies of Government and the Making of Subjects. New Ecologies for the Twenty-First Century. Durham: Duke University Press.
Argyrou, V.
2005   The Logic of Environmentalism: Anthropology, Ecology and Postcoloniality. New York: Berghahn Books.
Berreman, G.D.

1978   Ecology, Demography and Domestic Strategies in the Western Himalayas. *Journal of Anthropological Research* 34(3): 326–368.

Berti, D.

2015   Gods' Rights vs Hydroelectric Projects. Environmental Conflicts and the Judicialization of Nature in India. In *The Human Person and Nature in Classical and Modern India*. R. Torella and G. Milanetti, eds. Pp. 111–129. Supplemento n°2 alla Rivista Degli Studi Orientali, n.s., vol. LXXXVIII.

Bruun, O. and Kalland, A.

1995   Asian Perceptions of Nature: A Critical Approach. Curzon.

Chakrabarty, D.

2009a   Dipesh Chakrabarty – Breaking the Wall of Two Cultures. Science and Humanities After Climate Change. In *Falling Walls Lectures*. http://www.falling-walls.com/videos/Dipesh-Chakrabarty--1225, accessed October 30, 2015.

2009b   The Climate of History: Four Theses. *Critical Inquiry* 35(2): 197–222.

CNN

2013   Pesticides Found in Deadly School Lunch in India. By Watkins, T. and Singh, S.H. for *CNN*. https://edition.cnn.com/2013/07/22/us/pesticides-found-in-deadly-school-lunch-in-india/index.html, accessed May 2017.

Cronon, W.

1996a   Introduction: In Search of Nature. In *Uncommon Ground: Rethinking the Human Place in Nature*. William Cronon, ed. Pp. 23–56. New York: W.W. Norton.

1996b   The Trouble with Wilderness; or, Getting Back to the Wrong Nature. In *Uncommon Ground: Rethinking the Human Place in Nature*. William Cronon, ed. Pp. 69–90. New York: W.W. Norton.

Cruikshank, J.

2001   Do Glaciers Listen? Local Knowledge, Colonial Encounters, and Social Imagination. *Arctic* 54(4): 377–393.

Deleuze, G.

1999   Foucault. London, UK: The Athlone Press.

Descola, P.

2013   Beyond Nature and Culture. London: The University of Chicago Press, Ltd. United States of America.

Dove, M.R.

1992   The Dialectical History of "Jungle" in Pakistan: An Examination of the Relationship Between Nature and Culture. *Journal of Anthropological Research* 48(3): 231–253.

Dubash, N.

2012   Handbook of Climate Change and India: Development, Politics and Governance. New Delhi: Oxford University Press.

European Climate Adaptation Platform

2015   Awareness Campaigns for Behavioural Change. https://climate-adapt.eea.europa.eu/metadata/adaptation-options/awareness-campaigns-for-behavioural-change, accessed May 21, 2018.

Foucault, M.

1995   The Archaeology of Knowledge. A. M. Sheridan Smith. Reprint. London and New York: Routledge.

Foucault, M. and Rabinow, P.

1984   The Foucault Reader. New York, USA: Pantheon Books.

GRID-Arendal

2020   GRID-Arendal. https://www.grida.no/, accessed June 2020.

Griessler, E. and Littig, B.

2005   Social Sustainability: A Catchword Between Political Pragmatism and Social Theory. *International Journal for Sustainable Development* 8(1/2): 65–79. Open Access pp. 1–2.

Gupta, A.

2000   Postcolonial Developments: Agriculture in the Making of Modern India. Durham, NC: Duke University Press.

Haberman, D.

2006   River of Love in an Age of Pollution, The Yamuna River of Northern India. Berkeley: University of California Press.

Haryana Government

2010   Mission Green Haryana. Haryana Govt. Environment Press Releases. http://haryanagovt.blogspot.com/2010/03/mission-green-haryana.html, accessed April 2017.

Hastrup, K.

2013   Scales of Attention in Fieldwork: Global Connections and Local Concerns in the Arctic. *Ethnography* 14(2): 145–164.

Heaven in Himachal

2020   Instagram Posts About. https://www.instagram.com/heaveninhimachal/?hl=nb/, accessed June 2020.

Huber, T. and Pedersen, P.

1997   Meteorological Knowledge and Environmental Ideas in Traditional and Modern Societies: The Case of Tibet. *The Journal of the Royal Anthropological Institute* 3(3): 577–597.

Hulme, M.

2009   Why We Disagree about Climate Change: Understanding Controversy, Inaction and Opportunity. 4th edition. Cambridge, UK and New York: Cambridge University Press.

Ingold, T.

1995   Globes and Spheres: The Topology of Environmentalism. In Environmentalism: The View from Anthropology, Kay Milton, ed. Pp. 31–42. London and New York: Routledge.

2000   The Perception of the Environment: Essays on Livelihood, Dwelling and Skill. Reissue. London and New York: Routledge.

2011   Being Alive: Essays on Movement, Knowledge and Description. London and New York: Routledge.

Knudsen, A.

2011   Logging the 'Frontier': Narratives of Deforestation in the Northern Borderlands of British India, c. 1850–1940. *Forum for Development Studies* 38(3): 299–319.

Lahsen, M

2009   A Science–Policy Interface in the Global South: The Politics of Carbon Sinks and Science in Brazil. *Climatic Change* 97: 339–372.

Leichenko, R. and O'Brien, K.

2019   Climate and Society. Transforming the Future. Cambridge, UK: Polity Press.

Leiserowitz, A. and Thaker, J.

2012   Climate Change in the Indian Mind. Yale Project on Climate Change Communication, Yale University.

Makarieva, A.M., Gorshkov, V.G., Sheil, D., Nobre, A.D. and Li, B.-L.
2013   Where Do Winds Come From? A New Theory on How Water Vapor Condensation Influences Atmospheric Pressure and Dynamics. *Atmospheric Chemistry and Physics* 13: 1039–1056. doi: https://doi.org/10.5194/acp-13-1039-2013.

Mawdsley, E.
2006   Hindu Nationalism, Neo-Traditionalism and Environmental Discourses in India. *Geoforum* 37(3): 380–390.

Milton, K.
2002   Environmentalism and Cultural Theory. Exploring the Role of Anthropology in Environmental Discourse. e-Library edition. London, USA and Canada: Routledge, Taylor & Francis.

Mornee Tramp
2016   Morni Hills. Blog by Amitabh Dhillon. http://www.hillsofmorni.com/, accessed December 6, 2016.

Nanda, M.
2004   Dharmic Ecology and the Neo-Pagan International: The Dangers of Religious Environmentalism in India. *Paper Presented at the 18th European Conference on Modern South Asian Studies*, Lunds University, Sweden.

Nelson, L.E.
1998   Introduction. In *Purifying the Earthly Body of God: Religion and Ecology in Hindu India*. Pp. 1–10. SUNY Series in Religious Studies, L.E. Nelson (ed.). Albany: State University of New York Press.

The Nobel Foundation
2014   The Nobel Peace Prize 2007. http://www.nobelprize.org/nobel_prizes/peace/laureates/2007/, accessed February 21, 2017.

Norgaard, K.M.
2011   Living in Denial: Climate Change, Emotions, and Everyday Life. Cambridge, MA and London, England: MIT Press.

Ranjan, A.
2019a   The Advent of Environmental Issues in India's Elections. Published Opinion in *The Wire*. https://thewire.in/environment/the-advent-of-environmental-issues-in-indias-elections, accessed May 2020.
2019b   Does the Environment Matter to Indian Voters? Published Opinion in *The Wire*. https://thewire.in/environment/does-the-environment-matter-to-indian-voters, accessed May 2020.

Rosa, H.
2019   Resonance. A Sociology of Our Relationship to the World. [2016]. Cambridge, UK: Polity Press.

Scott, J.C.
1998   Seeing Like a State: How Certain Schemes to Improve the Human Condition Have Failed. New Haven and London: Yale University Press.

Tsing, A.L.
2005   Friction: An Ethnography of Global Connection. Princeton, NJ and Oxford: Princeton University Press.

Turnbull, D.
2005   Masons, Tricksters and Cartographers: Comparative Studies in the Sociology of Scientific and Indigenous Knowledge. London: Routledge.

6 A dance of global warming

Garmī

The hot summer season is the longest in Rani Mājri and consists of almost three months: the last half of Ćaīt, which starts in in early April, the month of Vaisākh and the month of Asāṛh in the middle of June.

There are fewer birds singing in the mornings in the hot season, and during the day few animals are heard; only the crickets make an inferno, playing their never-ending tune all through the sweltering night. The flowers have withered; the summer grass is turning yellow and heavy with seeds. This is the time to harvest the chickpeas and the all-important wheat.

During this major harvest, the houses in Rani Mājri are almost empty. Everyone is out in the fields at the same time – the landowners, the families and the S.C.s: everyone contributes for a share of the crop. By early morning, before breakfast, most adults are already on the field. The housewife will be the last to leave, after tending to the cattle and making breakfast to bring to everyone else. For hours, hunched down amidst the wheat, one sees nothing but rows of golden yellow and one's nearest harvesting fellows. Working back-to-back inside the yellow forest, cutting the wheat with the small hand-held scythe feels like being inside of a threshing machine. Dust flies up, scarves are tied around the face, the heat is intense. Short water-and-tea breaks are welcomed. The old and reused plastic bottles are kept in the shade of the few trees left on the fields, where everyone gathers to rest. The precious water is shared in that impressive technique where the bottle never touches the mouth so that it can be drunk by all. The mood is light and relaxed; conversation flows casually. The S.C.s who work for the landowning households also join in the conversation, although they are seated a few feet away from the landowners, outside the brittle shade of the trees.

Skin turns to a deep brown, almost black, where it is exposed to the sun. In some, the sun causes sunspots, including light markings on the faces of the older women. The younger ones are bothered by more acne, including small pimples and heat rashes, than usual. The young girls make facemasks of curd and chickpea flour to sooth their skin reactions whilst longing for the up-market skin-bleaching creams they see on television. The thick skin on heels cracks in *the dry air, keeping men and women awake at night with pain. Diarrhoea, vomiting and fever are other afflictions often attributed to having "worked in the sun". The hot, early summer days are long. No one goes home to the house before darkness falls, and dinners are eaten as late as 8 or 9 pm. During dinner, there's little talk. People go straight to sleep.*

As two weeks of intense labour pass, the last of the harvest is completed. To thank Panch Pīr, women dress up in beautiful attire and walk down to the fields with trays – an offering to the deity for the harvest. In high summer, it is so hot the air stirs. The wind, which otherwise would have been welcoming, feels like a hairdryer on the skin. No sweat is ever left to cool the body: in the dry heat it vaporises instantly, leaving only a prickly sensation and a salty residue. At midday people withdraw indoors to sleep and rest, preferably under a fan, but the village has more frequent and longer power-cuts now as the demand for air conditioning rises in the cities of the plains. The electricity grid repeatedly reaches its capacity, working to keep the temperature down in malls, offices and private homes.

*The heat that now develops is so intense that all outdoor work is lowered to a minimum. Tending to the buffaloes, cooking, eating and washing – all is done as early as possible in the morning to avoid the worst of the heat. The water tank for the cattle has more algae than water now, and there is no water in the kuhl for most of the day, except for when the tanks are open for some laundry and the kitchen gardens. The fields look dead – hard as stone, cracking up, dry as a desert. There are few vegetables now: some green bell peppers and some bottle gourd (*ghīa)*. Otherwise it is lentil soup (*dāl) every day, with sour fermented milk (*lassi) to drink.*

Everyone talks a lot about the forthcoming rain these days, not just the landholders – everyone wants a respite from the heat. The rain is important for everyone "for everything, everything needs water", as the shopkeeper tells me. In the middle of Asārh, in early June, something in the air changes. A thunderclap in the distance. Quickly, it gets darker, so fast it's eerie. The temperature plunges. Strong gusts of wind carry dark clouds over the village – clouds that have previously drifted tauntingly past it – and several lightning bolts strike ground close enough to smell. The clouds release a short and intense shower of rain. The village literally wakes up from slumber. Where people just hours ago had been weighed down by the intensity of the sun and the extreme, dry heat of summer, the winds and clouds that arrive from the plains cause hectic activity. Old ladies place their plastic chairs outside in the rain, and young girls dance in the water cascading from the roofs. As suddenly as they appeared, the clouds disappear, and the skies again clear for the scorching sun. These small and sudden showers at the end of summer fall irregularly for a few weeks. They will make the soil moist and ready for ploughing, and there are many preparations to be made before the "real" rains begin.

*In the midday sunshine, "work goes on" (*kām caltā hai)*. Rubbish, dead leaves and debris are swept away or burned to make everything clean (*sāf safāī)*, to let the water flow freely in the kuhl and to deprive mosquitoes and* *flies of places to lay eggs. Before the tilling can commence, the soil must be picked clean of stones, twigs and pebbles. Despite the discomfort of the humidity, literally everyone able is picking up stones that have appeared on the soil surface since last growing season.*[1] *One by one, stones are removed by hand and carried away in baskets. The larger ones are used to maintain the low, dry-stone fences that separate the fields and plots; others are thrown in small piles by the road. Broken fences and withered stalks of old tomato plants or grains are removed and sorted, preparing everything for* barsāt, *the monsoon season, to arrive.*

....

In the dry summer, the rain was anticipated in Rani Mājri with anxiety. As we awaited the monsoon season, I often asked whether the villagers thought it would come "on time" or if it had been "timelier" before. The adults and the elderly gave quite coherent replies. There was no way of knowing whether the monsoon would come "on time" because the right time was up to Bhagvān (God). The Rajput widow Lalita, around 80 years old at the time, would say, "When the rain comes is not in our hands. That, birth and death – these are three things we cannot do anything about – this is not for us to know, they are for Bhagvān to know". The hills around Rani Mājri carry the scars of numerous landslides, a continuous reminder that the ground beneath your feet can shift to your disadvantage at any time. After the 1990s, the farmers agreed that the monsoons had been particularly irregular and difficult to foresee. A weak monsoon season could cause depleting droughts and critical food-crop failures; a sudden and erratic monsoon season could cause severe landslides in the porous soil, flooding and havoc.

The monsoon season of July 2013 was of the erratic kind and it did cause havoc. In the eastern state of Uttarakhand, in the upper Himalayan regions, disaster occurred. A cloud-break lasting for several days, combined with the rapid melting of snow and ice in the mountains. The sudden discharge caused a disastrous flood in a form of a cascading torrent throughout its surge downhill, which the State's Chief Minister, Vijay Bahuguna, called "the Himalayan Tsunami" (B.B.C. 2013). Kedarnath, thought to be the abode of Shiva, was badly hit (Whitmore 2018). This mountainous region is normally sparsely populated, but in May and June every year these hills are crowded with visitors following the pilgrimage of the "Chota Char Dham"[2] (the Four Small Abodes) – popular religious sites for both Sikhs and Hindus. The flood thus hit at a particularly disastrous time as it obliterated several villages, displaced 100,000 people and caused the deaths of several thousands more (the numbers reported varied between 4,000 and 6,000 people dead[3]). The monsoon season had begun with ferociousness.

In the following weeks, red, fat-font numbers communicated the number of dead, missing or injured on the local television news, accompanied by bird's-eye helicopter images of the difficult rescue operations being carried out by military forces. As with any disaster of such magnitude, people would talk about why and how this could happen, and the quest for answers coloured both dinner conversations and the national media in the days and weeks that followed. The major news publishers flagged global warming and climate change, but the scientific community seemed divided on where to place the blame. Many argued that, even though the extreme snow melt in the Himalayas had been caused by warmer seasonal temperatures and an increasingly erratic monsoon season, the devastating outcome of the flood was related just as much to the government's (mis)management of forest and hill ecology. Social scientist Mathur (2015) even argued that the over-emphasis on climate change as the culprit for the incident by the Indian state was done with intent to shroud other and more harmful "human-induced policies and practices" (Mathur 2015:102). A hybrid story of the "unnatural" flood thus emerged, caused when state (mis)management and global warming came together in a lethal combination. One example of how the story was interpreted as a "hybrid" incident was found in the English newspaper *The Hindu*, where the environmental activist and director of People's Science Institute, Ravi Chopra, argued that the floods were far from a "freak accident" but, rather, was a sign of what was to come. Chopra wrote:

> Several reports from the Intergovernmental Panel on Climate Change (IPCC) have repeatedly warned that extreme weather incidents will become more frequent with global warming. ... With utter disregard for the State's mountain character and its delicate ecosystems, successive government has blindly pushed roads, dams, tunnels, bridges and unsafe buildings even in the most fragile regions. Last week's flood has sounded an alarm bell. To pursue development without concern for the fragile Himalayan environment is to invite disaster.
>
> (Chopra 2013)

The article pointed to issues of development, over-consumption, environmental deterioration and faulty state governance. Many of those I talked with in the city of Chandigarh, as well as in the village of Rani Mājri, offered similar explanations as the most likely cause of the "Himalayan Tsunami". The employees of the regional Development Office; my Chandigarh-based taxi-driver; the shopkeepers in the Chandigarh Sector 15 market; my Chandigarh landlords; and the neighbours, friends and acquaintances I made during our time in the city were largely unanimous: global warming and the short-sightedness of the government in building roads, dams and hotels where there should have been none were behind the tragedy. However, another dimension was also added to this hybrid story: namely the role played by Shiva.

During the media coverage of the incident, two images circulated in the news more than others. One depicted a flooded Kedarnath with buildings collapsing into the surging water, only one left standing– a massive Shiva temple. The other depicted a large Shiva statue, almost submerged in the floodwater of Rishikesh, with the head standing above water. Both images became iconic for the disaster and, for many, seemed to embody the mighty force that Hinduism ascribes to lord Shiva, the same deity that was seen by many Hindus as having initiated the Uttarakhand flood. One survivor described the disaster as follows: "It was like Lord Shiva doing his Tāṇḍav" (*Global Post* 2013), referring to the dance where Shiva comes out of his ascetic state and destroys in order for creation to take place (Fuller 2004:36). These religious undertones added yet another dimension to the way the disaster was responded to in Rani Mājri.

When news reached the village, people were obviously concerned, and the relative proximity of the site also meant that many had friends or relatives affected by the flood. I was no less concerned myself, having the local landslide fresh in my mind, and would often ask people, "Do you think something like that can happen here?" and "Why do you think it happened?" The answers I got varied, from those who were so concerned that they slept poorly to those who assured me that I need not worry. But no one considered a flash flood of such magnitude to be a "normal" event.

Prakash and his younger brother Bikram were watching the "News Nation" morning channel a few days after the incident. The programme was showing snippets of post-flood reportages, with dramatic replays of houses collapsing into a frothing river, desperate people being stuck on hillsides, helicopter rescue operations, etc. Prompted by the dramatic images on screen, I asked whether these massive landslides and floods were normal for the monsoon season. Prakash explained to me that indeed they were, but these consequences were not. As I knew from earlier conversations, Prakash was vaguely familiar with the term "global warming", so I asked him if he thought the disaster had been caused by this. He shook his head confidently. No, this had originated from the hands of God. I was a bit surprised by his response. Prakash had earlier demonstrated a tendency to distance himself strongly from what he considered to be "nonsensical speculations" and rumours – be they about residue from pesticide in locally produced foods or stories of ghosts and spirits. In fact, he said he knew the reason why Shiva had become upset too: an old Goddess' temple had been demolished at the flood site (or removed – he was not sure). A new one had been built on a different site, and this relocation had angered the lord Shiva. In his anger, he had sent the flood as a punishment for the wrongdoing. I asked Prakash where he had heard this story, and he said he had got the information from Jammu, where his oldest daughter, Padma, had gone to complete an exam. The family she stayed with there had family living close to the flood, so they knew all about it.

That a relocation of a temple was to blame was not a unique explanation in Rani Mājri. I encountered a few early news reports on the incident where explanations like the one I heard from Prakash were given. A medical relief teaming working on-site, for example, told a reporter from the *Global Post* that:

For the locals, the floods were more than just a natural disaster. A priest told us that the government had angered the gods and brought this disaster upon the people. According to legend, he said, the Char Dam pilgrims were protected by Dhari devī (an avatar for the Hindu Goddess, Kali Mata). However, the idol had been shifted from its original place in view of the Hydel Power Project. This angered the devī and Lord Shiva, who's one form is Kedarnath, and a few hours later there was cloudburst and floods, he said. He added that a king in the 1880's had made a similar attempt, which with similar results. Many locals and religious leaders had strongly opposed the shifting of the idol.

(Global Post 2013)

Shiva and his anger over the relocated temple continued to colour the explanations provided in the days following the incident, but after a week's time, the reason for the flood had become more abstract. When explaining it, people would more often point to a general decay of contemporary society. A few weeks after the incident, two Rajput women in their late thirties became quite agitated when talking about the flood and what it meant. According to them, it was the moral corruption of the world today that had made Shiva so angry. "There is sin in the world, filth/dirt (*gandagī*) has spread in it!" they exclaimed with passion. That contemporary society affected the minds of the young generation negatively was not an uncommon point of view, and the argument fell in line with the familiar complaints that today's youth no longer respect the elderly and chase their own individual successes before thinking about their families or the gods (see also Berti 2012). Shiva sees this and had sent the flood as a punishment or warning. The two Rajput women emphasised especially the fact that Shiva had killed his own devotees as this underscored the sincerity and severity of his anger – they, of all people, ought to have known better. In 2016, when three years had passed, I made a short visit to Rani Mājri The people I talked to about the incident in Uttarakhand still seemed to think that it had been caused by Shiva and the wrongdoing of men. One middle-aged Rajput couple told me that, in fact, Shiva was upset that people did not drink the river's water anymore but instead bought tapped and bottled water. This should be seen in context with recent reports of booming water bottle sales (see, for example, *The Economic Times* 2016).

These religiously informed explanations were not unequivocal, however. As I had also picked up in Chandigarh, in Rani Mājri people would also blame the stupidity and short-sightedness of men, particularly politicians, planners and developers in the region. One of the more influential and wealthy Rajput farmers of the village, upon reflecting on the incident, sighed that "people never learn". Why they had built their hotels "and all" on an old riverbed, prone to floods, he could not understand. "With heavy rainfall, the water will always find its old course and flood the place", he said. As the young Rajput woman from Khot who I talked to would argue, the topography of Uttarakhand was also more deforested, and the hills were even steeper than those of Rani Mājri, meaning that the village would probably be safe from these kinds of flash floods. When I asked her whether the flood had happened because Shiva was upset, she answered, "No, I don't think so... but even if that was the case, Bhagvān is so very far away from here".[4]

Common among these religious perceptions is a connection in which the environment, including the weather and the warming of the globe, ties up with notions of "right" and "wrong" behaviour. In Hinduism, this would be explained by the notion of *dharma* and *pāp*. *Dharma* refers to adhering to a particular moral order of the universe; "that which is to be kept", "religious observance", or "prescribed code of conduct" (Platt 2020), whilst *pāp* can be translated as sin. According to Wadley and Derr's (1989) study of the responses to a village fire in Karimpur, North India, in 1984, all actions in life are categorised as either sin (*pāp*) or merit (*puṇya*). It is *dharma* that defines a person's actions (*karma*) as good or bad (Wadley and Derr 1989). Are the irregular weather patterns, the heating of the globe and the contemporary changes that the villagers observe perceived as being part of a God's play? If so, what does that indicate?

Just weeks after the flood in Uttarakhand, a smaller landslide occurred in Rani Mājri, as described in the prologue. This incident struck closer to home. I was told that the deity of rain, Khwaja Pīr, had neglected to protect the house, but whether the villagers thought this was because of anger or just him haphazardly glancing past the village, I do not know. Unlike the Uttarakhand flood, I could not make out a specific act or moral breach by the family or village that had initiated the heavy rain. Instead, Bhagwati, Prakash's mother, stated that this was a part of their family's fate[5] (*qismat*) – this misfortune had been bound to happen. In Rani Mājri, the notion of *qismat* was invoked in several contexts, but most often as an explanation for misfortune, as in the local landslide, mentioned above. *Qismat* is part of the natural order, and by adhering to *dharma*, one can be given merit decisive for one's *qismat*. In Hinduism, one's fate is written at birth on one's forehead. But *karma* and its relationship to *qismat* is not as "fixed" or "set" as the concept of "fate" in English implies. When, for example, I asked Nirmala what to do if my *qismat* was to have only sons, she advised me to try this: "Tell the *devī* that if she gives you a girl, you will visit her temple and give 12 or 101 Rupees or whatever you have every time, and take fast so and so often – then maybe". If I approached the *devī* with enough devotion, she said, my *qismat* could be adjusted. As described in Chapter 4, deities and humans do not stand in passive relation to each other, but rather respond to each other's choice of action. In neither of the cases where the notion of *qismat* was invoked did the individuals or groups afflicted by misfortune passively accept destiny; neither could that destiny be augmented by ritual alone. If anything, destiny was actively manipulated by all options available, be they based upon human or upon divine relationships. But when all known remedies fail, one is forced to make "adjustments" (*adjust karna*) to the new scenario so as to mitigate the consequences.

This also bestows Hindus with the agency to continuously improve (or worsen) their current (and next) life, far from the "oriental" cliché of Hindu cyclicity (Good 2000). In my view, this removes the deterministic notion from Hindu cosmology: the idea of Hindus believing in a fated, as in fatalistic, process of environmental decay (as suggested by Nelson 1998). Instead, the processes leading to global warming are neither unavoidable nor fatalistic but, rather, a matter of adjusting oneself to obstacles and challenges in the "right" or "wrong" manner.

Environmental retribution for the "wrong" progress

Explaining floods, fires or extreme weather conditions as the result of human immorality is far from a new concept, nor is it a thing of the past. Orlove et al. (2009) noted that "[i]n many communities, weather and climate are understood as part of a universe infused with spiritual significance. Perturbations are often interpreted in terms of violation of religious, moral, and social norms" (Orlove et al. 2009:97). In Tibet, for example, Huber and Pedersen (1997) found the environment to be perceived as embodied in several social institutions and practices, making environmental change appear in a "moral climate" (ibid.:584–588). In Himachal Pradesh, Berti (2012, 2015) noted in her analysis of court cases regarding land rights and judicial practice that both individual and collective sin are thought to cause environmental retribution from a deity and that it does, in some cases, hold up in the courts of law. In Rajasthan, North India, Gold (1998, 1999) found that the villagers of Ghatiyali connected sociality and morality to the weather, and that, when villagers vocalised concerns about deforestation and the deterioration of rain and wildlife, they related it explicitly to a degradation of emotional bonds (Gold 1998:177). She also noted that, even if stories about the changing environment conflicted at times, "[p]eople portrayed interlocked changes in landscape, agriculture, society, religion and morality" (Gold 1998:166) with a causality that was neither singular nor linear – rather it was "fundamentally ecological in its sensitivity to the web-like interconnectedness of concurrent transformations" (ibid.). In fact, as people of Ghatiyali and Rani Mājri ate different foods, worked differently and behaved differently towards each other and their deities, the environment – as *vatavaran* – changed in retribution for the "wrong" kind of progress.

These observations are not confined to Asia or "Eastern" traditions. Rudiak-Gould (2012, 2013) looked at the idea of climate change as it unfolds in Ujae, in the Marshall Islands. There, "everyone" has heard about climate change and would generally rank it as the fifth-greatest concern in their lives, only surpassed by economic hardship, changing lifestyles, population growth, diabetes and other health problems (Rudiak-Gould 2013:90). In Ujae, however, many believed in a Christian God who was in control of the weather. Scientists were also believed by many, Rudiak-Gould found, but, as with all predictions of the future, there was no perfect certainty to be found either here or there (ibid.:69). Interestingly, Rudiak-Gould noted, Marshall Islanders were found blaming themselves for climate change, through "in-group" blaming and decline narratives. Central to this "in-group" blaming was how they related certain patterns of behaviour and consumption to ways of living associated with "modernity". When the Marshall Islanders talked about climate change, they talked about "oktak in mejatoto" or changes in the "mejatoto" (Rudiak-Gould 2012:47). "Mejatoto" does not correspond to "climate" or "weather" directly; rather, it encompass the sky, atmosphere, space, climate and weather – even the cosmos in general, which also includes the social sphere. Rudiak-Gould thus suggested that, as "mejatoto" can be said to encompass changes to "almost anything" from the past (ibid.:49), the Ujae people's perception of climate change is infused with a notion that there has been a "wrong" kind of "progress". He argues that modernity, to the Marshall Islanders, is like a "trickster". In its seductive notion of "progress", "modernity" lures the people of Ujae into a web of consumption and a practice of "living by money" (Rudiak-Gould 2013:161). The feeling of guilt thus comes from knowing that they could have let it be, but it is they who continue to consume in this way – because of the positive and useful developments and artefacts that come along with "progress" (ibid.:34). The "modernity as trickster" narrative fits uncannily well, Rudiak-Gould found, with a belief in climate change because "traditionalism" in the Pacific has much in common with "environmentalism". Both narrate a fall from a distant past "to a corrupted present and ruinous future" (Rudiak-Gould 2013:92).

Rasmussen (2016), in his ethnography of Catholic villagers in Recuay, in the Andean foothills of Peru, noted the same tendency of the environment (through God) to exact retribution because of unwanted "progress". In Recuay, people's livelihoods are sensitive to changes in precipitation and temperature, and local life revolves around a stable water supply from the rivers, where melting water runs from retreating glaciers. The Atoq Huacanca river, on which their livelihoods depend, has recently turned into a "rebel" or a "Diablo River" (devil river) – "a symptom of far larger things gone astray" (Rasmussen 2016:9). In Alaska, similar observations appear. Cruikshank (2001) looked at how glaciers figure into the oral traditions of the Tinglit of Yukon and revealed how "sentient" glaciers are to the Tinglit – they perceive them to listen, pay attention and respond to human behaviour (Cruikshank 2001:378). She argues that the Tinglit not only perceive a changing climate through shifting glaciers but also perceive changes to "everything", with their meanings "enmeshed in social worlds" (ibid.:382). We see here yet another example of the climate change process happening as a sort of environmental retribution for human behaviour – as a harsh comment on social life, morality, values and priorities in the time of a Janus-faced modernity.

It would also be wrong to draw only upon examples from small villages and places peripheral to the industrialised and influential cities and countries of the world. Similar religiously infused perceptions of contemporary climate change are everywhere. The Norwegian *Imam* Ibrahim Saidy, for example, had support from the Islamic Board of Norway when he called for a *"green jihad"* to make people aware of climate change. His use of "jihad" was a little untimely, perhaps, but it helped his plea go truly viral, especially since the story was picked up by Fox News in the U.S. (*Fox News* 2015). The Catholic Church, too, is devoted to raising awareness of climate change. In 2015, Pope Francis wrote a critical and very engaged, 184-page encyclical entitled "Laudato Si" (meaning "Praise to You"), expressing his concerns for planet Earth and calling for awareness of climate change issues from the followers of the Catholicism (Pope Francis 2015). Even cross-faith initiatives are forming. In 1996, the Alliance for Religions and Conservation (A.R.C.)[6] was formed, and, in 2007, a symposium was held by the NGO "Religion, Science and Environment" in Greenland. Here, 200 priests from the Christian, Jewish and Islamic faiths; scientists; theologians; and government officials gathered to discuss the changing environment of the Arctic and how humanity should respond in "responsible stewardship of the earth" (Pope Benedict XVI 2007).

With climate change both reviving traditional environmental movements and initiating new movements that draw on both left- and right-wing ideologies, it has become culturally attributed to "misguided" human behaviour, even outside the sphere of religion. One does not have to look much further than the speech Swedish climate activist Greta Thunberg gave at the UN's Climate Action Summit in New York in 2019:

> You have stolen my dreams and my childhood with your empty words. And yet I'm one of the lucky ones. People are suffering. People are dying. Entire ecosystems are collapsing. We are in the beginning of a mass extinction, and all you can talk about is money and fairy tales of eternal economic growth. How dare you!
> (Thunberg, 2019 cited in Encyclopaedia Britannica 2020)

Thunberg's speech epitomised the strong sentiments about the misguided behaviour of humans towards their environs, evoking sentiments of hope, anger and injustice amongst those who agree with her as well as those who don't.

This recasts the story in which the *whys* and the *hows* of global warming as a social and cultural force comment upon the changes we all experience. Perhaps it also aligns with what Whitmore (2018) calls "eco-social complexity", a tool used to envision that interrelatedness of humans and their surroundings. In line with Whitmore, the recurrent earthquakes, landslides and floods in the Himalayas act as a forceful reminder that "eco-sociality is a basic feature of human existence" (Whitmore 2018:23). When climate change causes seasons and rhythms to shift, not only in the Himalayas but across the globe, it is a reminder that awakens the notion of our dependence on the surroundings in which we dwell. It might also be an example of what Tsing (2005, 2010) called the practice of "worlding" – the uncomfortable but constant shifting between claiming and refusing context. Worlding, to Tsing, is "the always experimental and partial, and often quite wrong, attribution of world-like characteristics to scenes of social encounter" (Tsing 2010:48) and, more importantly, something all people do, all the time. Upon acknowledging that our perception of the world is always situated and accepting that these processes are perceived to be part of our social worlds, we must ask ourselves: why does it matter?

On reductionism and disempowerment

In the previous chapter, I indicated that climate change takes place within a discourse, or several discourses in fact, with equally universalist claims of what kind of progress is a "good" solution to the crisis, be it carbon capture, less consumption or green energy. Hulme (2009, 2017) when discussing the apparent disagreement about climate change in the "West", noted how different ideas exist about the cause and the drivers of the change, and how these ideas become infused with cultural meaning (2009). To make sense of the changes in climate, people draw upon their repertoire of knowledge, myths and beliefs, helping them understand the *whys* and *hows* in relation to us, our behaviour and our responsibilities (Hulme 2009); through this, climate change becomes intrinsically social. However, if the idea of climate change is part of a discourse that in its universality muddles its archaeology and its cultural specificity, the consensus that science aims to achieve on the subject "is problematic at best" (Hulme 2017:33, 37).

The "gap" described in Chapter 3 now appears as something other, or more than a transmission of scientific words and concepts; rather, it is a "gap" in perception of the *whys* and the *hows* of global warming and climate change. Hastrup (2013) wrote, of such gaps, that they might have potentially large implications. In her example, the hunters of High Arctic Greenland, where a strong scientific presence has been focussing on climate change for years, are encouraged to share their perceptions of climate change. Hastrup noted that, once weather variability is perceived as an effect of a scientific process by the hunters living there, "it implies an unknown agency in the social" (Hastrup 2013:158). The scientific explanation of climate change does not replace former perceptions, but rather enfolds within the perception the hunters might have had on those changes before. In Rani Mājri, religious and ideological conceptualisations of the workings of the world seamlessly mesh with people's intimate knowledge of deforestation, pollution and corruption. The coexistence of Gods and Goddesses with the very material and practical perceptions of geography and topography, not only regarding this incident but in the practicalities of everyday life, was striking. It made me wonder whether the two "narratives" about the flood could be argued to have been an example of what Ramanujan (1989) calls "compartmentalising". The concept is derived from psychology, and Ramanujan used it to describe how the Indian mind works to avoid the "dissonance" that might occur between "worldviews": for example when Indian Hindus learn Western science, business or technological practices. However, "compartmentalising" knowledge implies that there are two kinds of knowledge-archetypes that appear to be essentially different from each other. As the people in Rani Mājri never seemed to make that separation, "compartmentalising" does not agree with the empirical realities on the ground, nor does the concept agree with the theories of knowledge being localised and situated. Instead, as their actions are shaped in a discourse on "sustainable development", the people of Rani Mājri, too, will forge new understandings of the processes they are witnessing, as I outlined in Chapter 5.

In the purely scientific perception of climate change, devoid of any cultural meaning and shrouded in scientific ideal of neutrality, climate change is amoral, and it is apolitical. Science, Cruikshank aptly says, is not so *"self-consciously aware"* (Cruikshank 2001:391, italics added). Instead, the discourse on climate change, in its amoral, apolitical approach, can be less about bad governance, discrimination, marginalisation or global injustice and can veil the economic or social circumstances that bring those about. If we do not address them, the grand narratives behind *why* climate change happens and why it is happening *now* are largely left up for ideological and ontological grabs. This is how the *whys* and the *hows* of climate change can easily be caught up in other, more subjugating discourses on how people should best develop, mitigate and conserve. Pressed to the point: the way the discourse around climate change is communicated as "the absolute truth" about how the world and its populations function could also be accused of "black-boxing", or diverting attention from, other more pressing issues of inequality, marginalisation and "mismanagement" in those places where the consequences of the climate change process will be particularly acute. This is not, I believe, a willed consequence of the murky practices between climate change and Sustainable Development policies. It could be as simple as the lack of time spent talking to people in those areas, causing lack of sensitivity to how those policies become stubborn practical realities. It could also be because it is easier and more effective for those with plans and agendas to stay within a paradigm where large groups of people become in need of – and, indeed, dependent on – carefully monitored development. Or it could be, as largely experienced in Rani Mājri, simply a lack of care. Nevertheless, if one continues to passively accept the need for "awareness" and "sustainable development" in the face of climate change, and never pause to ask why, people on the frontlines of that climate change run the risk of becoming just as fragile, powerless and sensitive as the hills in which they make their living. In that, the climate change discourse can appear reductionist and can be boiled down to a kind of *climate determinism* (Hulme 2011).

I do not find the effort to enlighten the villagers about global warming processes that they, willingly or not, partake in to be all negative. Nor do I want to argue that all aspects of "progress" are awry. To the contrary, I believe people need to be made aware of what they do not already know as they should be able to make informed decisions concerning their own lives. Knowing how to deal with new challenges, such as those associated with climate change, requires input from new sources, communicated through global networks and experts. For that, diffusion of information, what we believe to be true and valid knowledge, is essential. Only by knowing one's options can one even begin to hope that people make the desired choice – which here can be summed up in the ambitious formulation of the 1987 Brundtland Commission on Sustainable Development as a "development that meets the needs of the present, without compromising the ability of future generations to meet their own needs" (Dubash 2012:8). What I do argue, however, is that we might want to reconsider what kind of perception of the world is inherent to the discourse of climate change and what it might do, not only for the way we perceive human agency in the face of climate change but for people's sense of self. Robbins (2020) suggests that as powerful discourses entwine with behavioural and ideological norms, they also govern "what people think, and who they think they are" (Robbins 2020:211).

In a case from Tuvalu, a small Polynesian state in the Pacific, Farbotko and Lazarus (2012) showed that these global discourses might, in fact, disempower those very people the global institutions tries to assist by redirecting the power and influence out of the hands of those affected by them. The people of Tuvalu resist being "climate change refugees", not only because they have always been a migrating people, but because this discursive shift largely ignore socio-economic realities. This process "reinforces the view that climate is a unilinear vector" (Farbotko and Lazarus 2012:384). An unintended consequence of climate change reductionism, then, is that the discourse around it produces an image of certain groups or cultures as passive victims of climate change. However, the people of Ujae, Yukon, Rani Mājri or Tuvalu are neither ignorant nor in denial – they perceive themselves as empowered, actively pursuing a future riddled with ambiguity and uncertainty. In that, the current climate change discourse runs the risk of disempowering these groups, in a reductionist manner (Hulme 2011).

Concluding remarks

What is a good life for you? About what do you worry, and why? The answers I got to these questions during my fieldwork in 2012/13 went much like this: a good life would be no longer having poor access to public goods, such as hospitals and higher education. A good life would be not having to choose which child to give the richest food and the best education to and which child to deprioritise. A good life would be no longer being treated unfairly by the courts and given lower market prices for their produce. It would be, for many, no longer suffering under local caste oppression or gender discrimination. It would be no longer worrying so much about becoming sick from the water they drink, the food they eat, the air they breathe. And it would be no longer being "backward" – a term used by the villagers themselves and by others, when explaining to me the differences they perceived between people in Rani Mājri and the "progressed" people of cities. But, at the turn of the 21st century, being heard amongst the millions of voices competing for attention is not a straightforward enterprise. Through numerous projects and initiatives, initially intended to help the villagers reach a level of understanding where their behaviour could be guided more easily into conserving their natural resources and increasing their production, climate change-initiated projects gained momentum and credibility for aligning themselves with the kind of knowledge that is valued most: the rational kind of knowledge, *science*. Those who mastered that language could participate in a discussion on how to define the *whys* and *hows* of climate change. Those not initiated into this discursive sphere are rendered helpless victims, backward and in need of enlightenment. Many marginalised voices, such as those of the S.C.s or the uneducated women of Rani Mājri, lack a space in which their concerns, fears and alternative solutions can be heard, tried and negotiated (as all claims of "right" and "wrong" must be). Even provided with that space, they do not feel that they are entitled to make an argument because they see themselves as too "backward" to have one. Left untangled, climate change as discourse risks deflating or exhausting the potential that people have to change their own lives with the strength of inner motivation, creativity and imagination, and erodes those relations that matter for those who have to make the most radical adjustments towards an uncertain future.

....

When I did my surveys, I would occasionally ask what could make village life better and what people hoped for in the future. Usually, the answer provided would be something like this one from Aparajita, a Rajput mother my own age, who said, "there should be a good harvest, in every season there should be rain[fall] at the [right] time, and ... a doctor's office in the vicinity".[7] With full stomachs and money in hand, her family could adjust better to both the potential ills and the potential benefits of the future. A beneficial environment, a predictable climate and the availability of someone who could assist her in a time of need – that was good enough for her future self, and she knew that. Deforestation and pollution and global warming would not be good for her future self, and she knew that too. Aparajita saw these processes as being closely linked because, to her, they were.

In this book, I have addressed how the people of Rani Mājri, as I observed them, were acutely aware of climate change. First, if Climate Change Awareness is knowing that global warming is happening; that the monsoon has become irregular; that the Himalayan ice caps are melting; and that the

Indian nation, as well as the farmer, must adjust even more, then the people of Rani Mājri were aware. If being environmentally aware is to know there are toxins or chemical residue in the food you produce and give your children, in the air which you breathe or in the water that you drink; that the forest should not be over-logged so it can hold the soil in place so as to draw rain closer and provide firewood in winter; or that waste stuck in the terrain is problematic, then the people of Rani Mājri were fully aware.

This kind of awareness, however, failed to express itself in practices and conversations with government employees or scientists who worked (most notably) through the Haryana State Government to change – or assist the change of – Rani Mājri into a "progressed" condition. Something seemed to shroud their vision, somehow. As I have tried to show in the preceding chapters, "awareness" appeared to depend on the degree to which people were considered acquainted with a supposedly apolitical, even amoral, scientific fact about climate. When the consequences of climate change were embedded in that very moral and political sociality, its effect was mitigated in practices directed at mending the relationship among between people and deities. It is here that the problem of climate change as a discourse lodged within a myth of scientific rationality comes in. If climate change is about how humanity is experiencing the effects of it, as caused by global warming, a process of changes to temperature and precipitation on the planet that is causing changes to seasonal rhythm and weather, and that is, in turn, affecting local flora and fauna, all actors involved were in fact equally "aware", whether local Shivalik villagers or not. Upon questioning the cause and effect of this process, however, the answers were so far apart that they appeared impossible to reconcile. They were not. Despite the versatility of our social lives around the world, our knowledge about the world as we know it is as practical and tangible, and as abstract and diffuse, as it is enduring and sensible, volatile and creative. Does that make it impossible to agree, as it were, on how to best mitigate the consequences of climate change? I think not.

Both for the Marshall Islanders of whom Moore speaks (see also Rudiak-Gould 2012, 2013) and for the people in Rani Mājri around whom this book revolves, changes in the climate are changes in how we relate to the world around us. The deterioration of the environment is a consequence of the deteriorating bonds between humans as part of a world where humans, Gods and Goddesses, demons, ghosts and spirits affect the past, the present and the future of a commonly inhabited world. As scientists we can choose to portray their views as radically different from the Western scientific consensus. But are the views really at odds? In this book, I argue that they might not be. The anthropologist Henrietta Moore, in her anniversary lecture at the Anthropological Institute in Oslo, Norway (2014), suggested advocating against radical alterity when approaching an understanding of how people around the world experience changes in social and environmental terms. This is a salient argument now, when our differences threaten to weaken prospects of solidarity and affinity. Climate change issues press this point by addressing how we in fact share the same world in a way that might force an unprecedented strain on these very bonds of affinity. To keep them, we should take Moore's advice and work towards being in relation with each other, and not cut down the spaces that we share (Moore 2014). But how to proceed?

The anthropologist Unni Wikan (1992) argues, "to translate across cultures we must be willing to forgo some precision for the sake of enhanced human solidarity" (Wikan 1992). Wikan used "resonance" as a concept to describe a kind of cultural harmonisation process taking place in the tacit communication "beyond the words", as she explains it (Wikan 1992). Resonance is a metaphor that I have found to be useful when attempting to let the climate change idea encompass both social and ecological processes.[8] Resonance, drawn from music theory, is an indication of sound being reverberated in an object, something that happens when an object is exposed to a frequency of sound, making it vibrate. Ingold (2000:199), and later also Krause (2013), used resonance to describe the body-environment relation, and, here, resonance indicates the act of making something harmonise the actions of the body or mind. However, this conception of resonance risks glancing past difference at the expense of similarity. Not denying that resonance is a good concept for stressing the "mutual in-tuning" between humans (Wikan 1992) or the "in-tuning" in human-environment relationships (Ingold 2000), my use of the word here needs to encompass power and chaos as well. My own use of "resonance" is thus better approached as a metaphor as *rich response*, where resonance also embraces contradiction and resistance. In this, I draw on Rosa (2019), but only to a degree. Rosa's concept of resonance is less of a willed action of "tuning in" and more a conceptualisation of a *qualitative* way of relating to the world – experiencing that something or someone responds to your being. This is a wider and more ambitious use of resonance than what I try to convey here. My use of it thus draws on all the above and indicates the cognitive process that we allow in our imagination to make other people's worlds "resonate" with our own. If we depart from that, I believe we will find not a rigid dichotomy between scientific thought and the ontologies of "others" but a medley, a creative juxtaposition, that allows for an infinite array of configurations of the world to coexist in addition to one we can talk about. Only then can we make space for the serious conversation across those culturally embedded differences of which Hastrup (2013) speaks and close those imaginary and real "gaps" of perception. In that, there is potential, and I hope that my ethnography can add to that realisation.

Notes

1 Pebbles and stones are known to have upward mobility in sandy soil. There are several geological reasons for this, including frost heave, water erosion, soil expansion and contradiction through drought and rain, etc. (Dregne 1977).

2 Most popular are the four temples representing the three major sectarian strands in Hinduism: in Badrinath, pilgrims visit the abode of Vishnu; in Kedarnath, the abode of Shiva; and in Gangotri and Yamunotri, the abodes of the Goddesses of Ganga and Yamuna.

3 The Indian State Government in 2014 reported 169 people died, and 4,021 people went missing (presumed dead) (Satendra et al. 2014).

4 Bhagvān bahut, bahut dūr yahāṁ se.

5 Qismat, from Urdu lit.: "fortune, luck, chance, fate, destiny" (Platts Dictionary 2020).

6 At the time of publishing, A.R.C is undergoing a transformation into The International Network for Conservation and Religion, supported by the W.W.F (A.R.C. 2019).

7 Ki acchī faṣal honā cāhiye - har mausam meṁ time meṁ bāriś honā cāhiye, aur... ḍākṭar kā office ās-pās meṁ hai.

8 Rosa (2019) draws heavily on Marxism, as he argues that it is through the process of becoming modern – for all the good it has brought us in health and individual freedom – that the world has become disenchanted and has left us feeling alienated. Alienation, then – the feeling that we have no real relation to our surroundings – is a classic feature of a modernist society, and resonance is suggested as alienation's 'other'.

References

Alliance of Religion and Conservation (A.R.C.)
2019   A new International Network for Conservation and Religion Announced. A.R.C. Web Page. http://www.arcworld.org/news.asp?pageID=914, accessed May 2019.
B.B.C.
2013   India Floods: Toll Rises after Bodies Found in Ganges. *BBC India.* https://www.bbc.com/news/world-asia-india-23001947, accessed September 2016.
Berti, D.
2012   Ritual Faults, Sins, and Legal Offences: A Discussion About Two Patterns of Justice in Contemporary India. In *Sins and Sinners Perspectives from Asian Religions.* Phyllis Granoff and Koichi Shinohara, eds. Pp.153–172. Leiden and The Netherlands: Brill. doi: 10.1163/9789004232006_009
2015   Gods' Rights vs Hydroelectric Projects. In *Environmental conflicts and the Judicialization of Nature in India.* Supplemento n°2 alla Rivista Degli Studi Orientali, n.s., vol. LXXXVIII, ('The Human Person and Nature in Classical and Modern India', R. Torella & G. Milanetti, eds.), Pp. 111–129. Sapienza: Universita di Roma.
Chopra, R.
2013   The Untold Story from Uttarakhand. Opinion in *The Hindu.* https://www.thehindu.com/opinion/lead/the-untold-story-from-uttarakhand/article4847166.ece, accessed May 2017.
Cruikshank, J.
2001   Do Glaciers Listen? Local Knowledge, Colonial Encounters, and Social Imagination. *Arctic* 54(4): 377–393.
Fox News
2015   Muslim and Christian Clerics Back Climate Activism with Music, Fasting. https://www.foxnews.com/world/muslim-and-christian-clerics-back-climate-activism-with-music-fasting, accessed May 3, 2018.
Dregne, H.

1977   Desertification of Arid Lands. *Economic Geography* 53(4), The Human Face of Desertification: 322–331.

Dubash, N.
2012   Handbook of Climate Change and India: Development, Politics and Governance. New Delhi: Oxford University Press.

Encyclopædia Britannica
2020   About: Gretha Thunberg. https://www.britannica.com/biography/Greta-Thunberg#ref1276549, accessed June 2020.

Farbotko, C. and Lazrus, H.
2012   The First Climate Refugees? Contesting Global Narratives of Climate Change in Tuvalu. *Global Environmental Change* 22(2): 382–390.

Fuller, C.J.
2004   The Camphor Flame: Popular Hinduism and Society in India. 2nd edition. Princeton, NJ: Princeton University Press.

Global Post
2013   June's Flash-Floods in Uttarakhand, India Leave Devastation Akin to an Inland Tsunami. Bajwa, M. and Boparai, H. https://www.pri.org/stories/2013-07-06/junes-flash-floods-uttarakand-india-leave-devastation-akin-inland-tsunami, accessed December 2016.

Gold, A.
1998   Sin and Rain: Moral Ecology in Rural North India. In *Purifying the Earthly Body of God: Religion and Ecology in Hindu India*. Lance E. Nelson, ed. Pp. 165–196. SUNY Series in Religious Studies. Albany: State University of New York Press.
1999   Abandoned Rituals: Knowledge, Time and Rethorics of Modernity in Rural India. In *Religion, Ritual and Royalty: Rajasthan Studies*. Narendra Singhi, ed. Pp. 262–275. Rawat Publications.

Good, A.
2000   Congealing Divinity: Time, Worship and Kinship in South Indian Hinduism. *The Journal of the Royal Anthropological Institute* 6(2): 273–292. Retrieved August 30, 2020, from http://www.jstor.org/stable/2660896.

Hastrup, K.
2013   Scales of Attention in Fieldwork: Global Connections and Local Concerns in the Arctic. *Ethnography* 14(2): 145–164.

Huber, T. and Pedersen, P.
1997   Meteorological Knowledge and Environmental Ideas in Traditional and Modern Societies: The Case of Tibet. *The Journal of the Royal Anthropological Institute* 3(3): 577–597.

Hulme, M.
2009   Why We Disagree about Climate Change: Understanding Controversy, Inaction and Opportunity. 4th Edition. Cambridge, UK and New York: Cambridge University Press.
2013   Exploring Climate Change through Science and in Society: An Anthology of Mike Hulme's Essays, Interviews and Speeches. London and New York: Routledge.
2017   Weathered: Cultures of Climate. London: SAGE Publications Ltd.

Ingold, T.
2000   The Perception of the Environment: Essays on Livelihood, Dwelling and Skill. Reissue. London and New York: Routledge.

Krause, F.

2013   Seasons as Rhythms on the Kemi River in Finnish Lapland. *Ethnos* 78(1): 23–46.

Mathur, N.

2015   It's a Conspiracy Theory and Climate Change. Of Beastly Encounters and Cervine Disappearances in Himalayan India. *Hau: Journal of Ethnographic Theory* 5(1): 87–111. doi: http://dx.doi.org/10.14318/hau5.1.005.

Moore, H.

2014   Recorded Lecture with Henrietta L. Moore at the University of Oslo. https://www.sv.uio.no/sai/om/aktuelt/sai-50/opptakmoore.html, accessed December 16, 2015.

Nelson, L.E.

1998   Introduction. In *Purifying the Earthly Body of God: Religion and Ecology in Hindu India*. Lance E. Nelson, ed. Pp. 1–10. SUNY Series in Religious Studies. Albany: State University of New York Press.

Orlove, B., Crane, T. and Roncoli, C.

2009   Fielding Climate Change in Cultural Anthropology. In *Anthropology and Climate Change: From Encounters to Actions*. Susan A. Crate and Mark Nuttall, eds. Pp. 87–115. Walnut Creek, CA: Left Coast Press Inc.

Platts Dictionary

2020   John T. Platts Dictionary. A dictionary of Urdu, Classical Hindi, and English. London: W.H. Allen & Co., 1884.

Pope Benedict XVI

2007   Letter to the Ecumenical Patriarch of Constantinople on the Occasion of the Seventh Symposium of the Religion, Science and the Environment Movement. https://w2.vatican.va/content/benedict-xvi/en/letters/2007/documents/hf_ben-xvi_let_20070901_symposium-environment.html, accessed May 22, 2017.

Pope Francis

2015   Encyclical Letter Laudato Si' Of The Holy Father Francis On Care For Our Common Home. http://www.vatican.va/content/francesco/en/encyclicals/documents/papa-francesco_20150524_enciclica-laudato-si.html, accessed May 2018.

Rasmussen, M. B.

2016   Unsettling Times Living with the Changing Horizons of the Peruvian Andes. *Latin American Perspectives* 43(4): 73–86.

Ramanujan, A. K.

1989   Is There an Indian Way of Thinking? An Informal Essay. *Contributions to Indian Sociology*, 23(1): 41–58. SAGE Publications.

Robbins, P.

2020   Political Ecology: A Critical Introduction, 3rd Edition. Chichester: Wiley-Blackwell.

Rosa, H.

2019   Resonance. A Sociology of Our Relationship to the World [2016]. Cambridge, UK: Polity Press.

Rudiak-Gould, P.

2012   Promiscuous Corroboration and Climate Change Translation: A Case Study from the Marshall Islands. *Global Environmental Change* 22(1): 46–54.

2013   Climate Change and Tradition in a Small Island State: The Rising Tide. Routledge Studies in Anthropology, 13. New York: Routledge, Taylor & Francis Group.

Satendra, D., Kumar, K.J. and Naik, V.K.
2014   India Disaster Report. National Institute of Disaster Management, Ministry of Home Affairs, Government of India. http://admin.indiaenvironmentportal.org.in/files/file/India%20Disaster%20Report%202013.pdf, accessed June 2020

The Economic Times
2016   Bottled Water Market Growing Faster than Carbonated Drinks in India, Mirroring Global Trend. By Bhushan, R. https://economictimes.indiatimes.com/industry/cons-products/food/bottled-water-market-growing-faster-than-carbonated-drinks-in-india-mirroring-global-trend/articleshow/53714636.cms?-from=mdr, accessed May 2017.

Tsing, A.L.
2005   Friction: An Ethnography of Global Connection. Princeton and Oxford: Princeton University Press.
2010   Worlding the Matsutake Diaspora. In *Experiments in Holism*. Ton Otto and Nils Bubandt, eds. Pp. 47–66. Wiley-Blackwell.

Wadley, S.S. and Derr, B.W.
1989   Eating Sins in Karimpur. *Contributions to Indian Sociology* 23: 1.

Whitmore, L.
2018   Mountain Water Rock God: Understanding Kedarnath in the Twenty-First Century. Oakland: University of California Press. doi: https://doi.org/10.1525/luminos.61.

Wikan, U.
1992   Beyond the Words: The Power of Resonance. *American Ethnologist* 19(3): 460–482.

Index

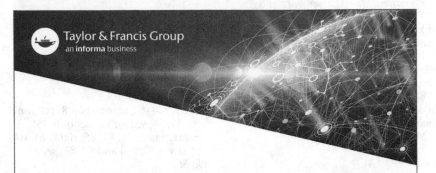

Printed in the United States
By Bookmasters